IN THE

BEGINNING...

WAS THE

COMMAND LINE

Also by
Neal Stephenson

CRYPTONOMICON
THE DIAMOND AGE
SNOW CRASH
ZODIAC

Neal Stephenson

IN THE BEGINNING... WAS THE COMMAND LINE

AVON BOOKS ◆ NEW YORK

AVON BOOKS, INC.
1350 Avenue of the Americas
New York, New York 10019

Copyright © 1999 by Neal Stephenson
Cover design by Raquel Jaramillo
Interior design by Elizabeth Glover
Published by arrangement with the author
ISBN: 0-380-81593-1
www.avonbooks.com

Library of Congress Cataloging in Publication Data:
Stephenson, Neal.
   In the beginning . . . was the command line / Neal Stephenson.
      p.   cm.
   1. Operating systems (Computers) I. Title.
QA76.76.063S7369   1999   99-40450
005.4'3—dc21   CIP

First Avon Books Trade Paperback Printing: November 1999

AVON TRADEMARK REG. U.S. PAT. OFF. AND IN OTHER COUNTRIES, MARCA
REGISTRADA, HECHO EN U.S.A.

Printed in the U.S.A.

OPM   10 9 8 7 6 5 4 3

IN THE
BEGINNING ...
WAS THE
COMMAND LINE

About twenty years ago Jobs and Wozniak, the founders of Apple, came up with the very strange idea of selling information-processing machines for use in the home. The business took off, and its founders made a lot of money and received the credit they deserved for being daring visionaries. But around the same time, Bill Gates and Paul Allen came up with an idea even stranger and more fantastical: selling computer operating systems. This was much weirder than the idea of Jobs and Wozniak. A computer at least had some sort of physical reality to it. It came in a box, you could open it up and plug it in and watch lights blink. An operating system had no tangible incarnation at all. It arrived on a disk, of course, but the disk was, in effect, nothing more than the box that the Operating System (OS) came in. The product itself was a very long string of ones and zeroes that, when properly installed and coddled, gave you the ability to manipulate other very long strings of ones and zeroes. Even those few who actually understood what a computer operating system was were apt to think of it as a fantastically arcane engineering prodigy, like a breeder reactor or a U-2 spy plane, and not something that could ever be (in the parlance of high tech) "productized."

Yet now the company that Gates and Allen founded is selling operating systems like Gillette sells razor blades. New releases of operating systems are launched as if they were Hollywood blockbusters, with celebrity endorsements, talk show appearances, and world tours. The market for them is vast enough that people worry about whether it has been monopolized by one company. Even the least technically minded people in our society now have at least a hazy idea of what operating systems do; what is more, they have strong opinions about their relative merits. It is commonly understood, even by technically unsophisticated computer users, that if you have a piece of software that works on your Macintosh, and you move it over onto a Windows machine, it will not run. That this would, in fact, be a laughable and idiotic mistake, like nailing horseshoes to the tires of a Buick.

A person who went into a coma before Microsoft was founded, and woke up now, could pick up this morning's *New York Times* and understand everything in it— almost:

. . . .

Item: the richest man in the world made his fortune from—what? Railways? Shipping? Oil? No, operating systems. Item: the Department of Justice has tackled Microsoft's supposed OS monopoly with legal tools that were

invented to restrain the power of nineteenth-century rob-
ber barons. Item: a woman friend of mine recently told
me that she'd broken off a (hitherto) stimulating exchange
of e-mail with a young man. At first he had seemed like
such an intelligent and interesting guy, she said, but then,
"he started going all PC-versus-Mac on me."

• • • • • • • • • • • • • • • • •

What the hell is going on here? And does the operating system business have a future, or only a past?
Here is my view, which is entirely subjective; but since
I have spent a fair amount of time not only using, but
programming, Macintoshes, Windows machines, Linux
boxes, and the BeOS, perhaps it is not so ill-informed
as to be completely worthless. This is a subjective essay,
more review than research paper, and so it might seem
unfair or biased compared to the technical reviews you
can find in PC magazines. But ever since the Mac came
out, our operating systems have been based on metaphors, and anything with metaphors in it is fair game as
far as I'm concerned.

MGBs, TANKS, AND BATMOBILES

Around the time that Jobs, Wozniak, Gates, and Allen were dreaming up these unlikely schemes, I was a teenager living in Ames, Iowa. One of my friends' dads had an old MGB sports car rusting away in his garage. Sometimes he would actually manage to get it running, and then he would take us for a spin around the block, with a memorable look of wild youthful exhilaration on his face; to his worried passengers, he was a madman, stalling and backfiring around Ames, Iowa, and eating the dust of rusty Gremlins and Pintos, but in his own mind he was Dustin Hoffman tooling across the Bay Bridge with the wind in his hair.

In retrospect, this was telling me two things about people's relationship to technology. One was that romance and image go a long way toward shaping their opinions. If you doubt it (and if you have a lot of spare time on your hands), just ask anyone who owns a Macintosh and who, on those grounds, imagines him- or herself to be a member of an oppressed minority group.

The other, somewhat subtler point, was that interface is very important. Sure, the MGB was a lousy car in almost every way that counted: balky, unreliable, under-powered. But it was fun to drive. It was responsive. Every pebble on the road was felt in the bones, every nuance in the pavement transmitted instantly to the driver's hands. He could listen to the engine and tell what was wrong with it. The steering responded immediately to commands from his hands. To us passengers it was a pointless exercise in going nowhere—about as interesting as peering over someone's shoulder while he punches numbers into a spreadsheet. But to the driver it was an experience. For a short time he was extending his body and his senses into a larger realm, and doing things that he couldn't do unassisted.

The analogy between cars and operating systems is not half bad, and so let me run with it for a moment, as a way of giving an executive summary of our situation today.

Imagine a crossroads where four competing auto dealerships are situated. One of them (Microsoft) is much, much bigger than the others. It started out years ago selling three-speed bicycles (MS-DOS); these were not perfect, but they worked, and when they broke you could easily fix them.

There was a competing bicycle dealership next door (Apple) that one day began selling motorized vehicles—expensive but attractively styled cars with their innards hermetically sealed, so that how they worked was something of a mystery.

The big dealership responded by rushing a moped up-grade kit (the original Windows) onto the market. This was a Rube Goldberg contraption that, when bolted onto a three-speed bicycle, enabled it to keep up, just barely, with Apple-cars. The users had to wear goggles and were always picking bugs out of their teeth while Apple owners sped along in hermetically sealed comfort, sneering out the windows. But the Micro-mopeds were cheap, and easy to fix compared with the Apple-cars, and their market share waxed.

Eventually the big dealership came out with a full-fledged car: a colossal station wagon (Windows 95). It had all the aesthetic appeal of a Soviet worker housing block, it leaked oil and blew gaskets, and it was an enormous success. A little later, they also came out with a hulking off-road vehicle intended for industrial users (Windows NT), which was no more beautiful than the station wagon and only a little more reliable.

Since then there has been a lot of noise and shouting, but little has changed. The smaller dealership continues to sell sleek Euro-styled sedans and to spend a lot of money on advertising campaigns. They have had GOING OUT OF BUSINESS! signs taped up in their windows for so long that they have gotten all yellow and curly. The big one keeps making bigger and bigger station wagons and ORVs.

On the other side of the road are two competitors that have come along more recently.

One of them (Be, Inc.) is selling fully operational Bat-mobiles (the BeOS). They are more beautiful and stylish even than the Euro-sedans, better designed, more technologically advanced, and at least as reliable as anything else on the market—and yet cheaper than the others.

With one exception, that is: Linux, which is right next door, and which is not a business at all. It's a bunch of RVs, yurts, tepees, and geodesic domes set up in a field and organized by consensus. The people who live there are making tanks. These are not old-fashioned, cast-iron Soviet tanks; these are more like the M1 tanks of the U.S. Army, made of space-age materials and jammed with sophisticated technology from one end to the other. But they are better than army tanks. They've been modified in such a way that they never, ever break down, are light and maneuverable enough to use on ordinary streets, and use no more fuel than a subcompact car. These tanks are being cranked out, on the spot, at a terrific pace, and a vast number of them are lined up along the edge of the road with keys in the ignition. Anyone who wants can simply climb into one and drive it away for free.

Customers come to this crossroads in throngs, day and night. Ninety percent of them go straight to the biggest dealership and buy station wagons or off-road vehicles. They do not even look at the other dealerships.

Of the remaining ten percent, most go and buy a sleek Euro-sedan, pausing only to turn up their noses at the philistines going to buy the station wagons and ORVs. If they even notice the people on the opposite side of the road, selling the cheaper, technically superior vehicles, these customers deride them as cranks and half-wits.

The Batmobile outlet sells a few vehicles to the occasional car nut who wants a second vehicle to go with his station wagon, but seems to accept, at least for now, that it's a fringe player.

The group giving away the free tanks only stays alive because it is staffed by volunteers, who are lined up at the edge of the street with bullhorns, trying to draw customers' attention to this incredible situation. A typical conversation goes something like this:

HACKER WITH BULLHORN: "Save your money! Accept one of our free tanks! It is invulnerable, and can drive across rocks and swamps at ninety miles an hour while getting a hundred miles to the gallon!"

PROSPECTIVE STATION WAGON BUYER: "I know what you say is true . . . but . . . er . . . I don't know how to maintain a tank!"

BULLHORN: "You don't know how to maintain a station wagon either!"

BUYER: "But this dealership has mechanics on staff. If something goes wrong with my station wagon, I can take a day off work, bring it here, and pay them to work on it while I sit in the waiting room for hours, listening to elevator music."

BULLHORN: "But if you accept one of our free tanks, we will send volunteers to your house to fix it for free while you sleep!"

BUYER: "Stay away from my house, you freak!"

BULLHORN: "But . . ."

BUYER: "Can't you see that everyone is buying station wagons?"

BIT-FLINGER

The connection between cars, and ways of interacting with computers, wouldn't have occurred to me at the time I was being taken for rides in that MGB. I had signed up to take a computer programming class at Ames High School. After a few introductory lectures, we students were granted admission into a tiny room containing a teletype, a telephone, and an old-fashioned modem consisting of a metal box with a pair of rubber cups on the top (note: many readers, making their way through that last sentence, probably felt an initial pang of dread that this essay was about to turn into a tedious, codgerly reminiscence about how tough we had it back in the old days; rest assured that I am actually positioning my pieces on the chessboard, as it were, in preparation to make a point about truly hip and up-to-the-minute topics like Open Source Software). The teletype was exactly the same sort of machine that had been used, for decades, to send and receive telegrams. It was basically a loud typewriter that could only produce UPPER-CASE LETTERS. Mounted to one side of it was a smaller machine with a long reel of paper tape on it, and a clear plastic hopper underneath.

In order to connect this device (which was not a computer at all) to the Iowa State University mainframe across town, you would pick up the phone, dial the computer's number, listen for strange noises, and then slam the handset down into the rubber cups. If your aim was true, one cup would wrap its neoprene lips around the earpiece and the other around the mouthpiece, consummating a kind of informational *soixante-neuf.* The teletype would shudder as it was possessed by the spirit of the distant mainframe and begin to hammer out cryptic messages.

Since computer time was a scarce resource, we used a sort of batch-processing technique. Before dialing the phone, we would turn on the tape puncher (a subsidiary machine bolted to the side of the teletype) and type in our programs. Each time we depressed a key, the teletype would bash out a letter on the paper in front of us, so we could read what we'd typed; but at the same time it would convert the letter into a set of eight binary digits, or bits, and punch a corresponding pattern of holes across the width of a paper tape. The tiny disks of paper knocked out of the tape would flutter down into the clear plastic hopper, which would slowly fill up with what can only be described as actual bits. On the last day of the school year, the smartest kid in the class (not me) jumped out from behind his desk and flung several quarts of these bits over the head of our teacher, like confetti, as a sort of semiaffectionate practical joke. The image of this man sitting there, gripped in the opening stages of an atavistic fight-or-flight reaction, with millions of bits (megabytes) sifting down out of his hair and into his nostrils and mouth, his face gradually turning purple as he built up to an explosion, is the single most memorable scene from my formal education.

Anyway, it will have been obvious that my interaction with the computer was of an extremely formal nature, being sharply divided up into different phases, viz.: (1) sitting at home with paper and pencil, miles and miles from any computer, I would think very, very hard about what I wanted the computer to do, and translate my intentions into a computer language—a series of alphanumeric symbols on a page. (2) I would carry this across a sort of informational *cordon sanitaire* (three miles of snowdrifts) to school and type those letters into a machine—not a computer—which would convert the symbols into binary numbers and record them visibly on a tape. (3) Then, through the rubber-cup modem, I would cause those numbers to be sent to the university mainframe, which would (4) do arithmetic on them and send different numbers back to the teletype. (5) The teletype would convert these numbers back into letters and hammer them out on a page, and (6) I, watching, would construe the letters as meaningful symbols.

The division of responsibilities implied by all of this is admirably clean: computers do arithmetic on bits of information. Humans construe the bits as meaningful symbols. But this distinction is now being blurred, or at least complicated, by the advent of modern operating systems that use, and frequently abuse, the power of metaphor to make computers accessible to a larger audience. Along the way—possibly because of those metaphors, which make an operating system a sort of work of art—people start to get emotional and grow attached to pieces of software in the same sort of way my friend's dad did to his MGB.

People who have only interacted with computers through graphical user interfaces such as the MacOS or Windows—which is to say, almost everyone nowadays who has ever used a computer—may have been startled, or at least bemused, to hear about the telegraph machine that I used to communicate with a computer in 1973. But there was, and is, a good reason for using this particular kind of technology. Human beings have various ways of communicating to each other, such as music, art, dance, and facial expressions, but some of these are more amenable than others to being expressed as strings of symbols. Written language is the easiest of all because, of course, it consists of strings of symbols to begin with. If the symbols happen to belong to a phonetic alphabet (as opposed to, say, ideograms), converting them into bits is a trivial procedure, and one that was nailed, technologically, in the early nineteenth century, with the introduction of Morse code and other forms of telegraphy.

We possessed a human/computer interface a hundred years before we had computers. When computers came into being around the time of the Second World War, humans, quite naturally, communicated with them by simply grafting them on to the already-existing techno-

logies for translating letters into bits and vice versa: teletypes and punch card machines.

These embodied two fundamentally different ap-
proaches to computing. When you were using cards,
you'd punch a whole stack of them and run them
through the reader all at once, which was called batch
processing. You could also do batch processing with a
teletype, as I have already described, by using the paper
tape reader, and we were certainly encouraged to use
this approach when I was in high school. But—though
efforts were made to keep us unaware of this—the teletype could do something that the card reader could not.
On the teletype, once the modem link was established,
you could just type in a line and hit the return key. The
teletype would send that line to the computer, which
might or might not respond with some lines of its own,
which the teletype would hammer out—producing, over
time, a transcript of your exchange with the machine.
This way of working did not even have a name at the
time, but when, much later, an alternative became available, it was retroactively dubbed the Command Line
Interface.

When I moved on to college, I did my computing in
large, stifling rooms where scores of students would sit
in front of slightly updated versions of the same machines
and write computer programs: these used dot-matrix
printing mechanisms, but were (from the computer's
point of view) identical to the old teletypes. By that
point, computers were better at time-sharing—that is,
mainframes were still mainframes, but they were better

at communicating with a large number of terminals at once. Consequently, it was no longer necessary to use batch processing. Card readers were shoved out into hallways and boiler rooms, and batch processing became a nerds-only kind of thing, and consequently took on a certain eldritch flavor among those of us who even knew it existed. We were all off the batch, and on the command line, interface now—my very first shift in operating system paradigms, if only I'd known it.

A huge stack of accordion-fold paper sat on the floor underneath each one of these glorified teletypes, and miles of paper shuddered through their platens. Almost all of this paper was thrown away or recycled without ever having been touched by ink—an ecological atrocity so glaring that those machines were soon replaced by video terminals—so-called glass teletypes—which were quieter and didn't waste paper. Again, though, from the computer's point of view, these were indistinguishable from World War II–era teletype machines. In effect we still used Victorian technology to communicate with computers until about 1984, when the Macintosh was introduced with its Graphical User Interface. Even after that, the command line continued to exist as an underlying stratum—a sort of brainstem reflex—of many modern computer systems all through the heyday of graphical user interfaces, or GUIs, as I will call them from now on.

GUIs

Now the first job that any coder needs to do when writing a new piece of software is to figure out how to take the information that is being worked with (in a graphics program, an image; in a spreadsheet, a grid of numbers) and turn it into a linear string of bytes. These strings of bytes are commonly called files or (somewhat more hiply) streams. They are to telegrams what modern humans are to Cro-Magnon man, which is to say, the same thing under a different name. All that you see on your computer screen—your Tomb Raider, your digitized voice mail messages, faxes, and word-processing documents written in thirty-seven different typefaces—is still, from the computer's point of view, just like telegrams, except much longer and demanding of more arithmetic.

The quickest way to get a taste of this is to fire up your web browser, visit a site on the Net, and then select the View/Document Source menu item. You will get a bunch of computer code that looks something like this:

In the Beginning . . .

```
<HTML>

<HEAD>

<TITLE>C R Y P T O N O M I C O N<TITLE>

</HEAD>

<BODY  BGCOLOR="#000000"  LINK="#996600"  ALINK=
"#FFFFFF" VLINK="#663300">

<MAP NAME="navtext">
   <AREA  SHAPE=RECT  HREF="praise.html"  COORDS="0,37,
84,55">
   <AREA  SHAPE=RECT  HREF="author.html"  COORDS="0,59,
137,75">
   <AREA SHAPE=RECT HREF="text.html" COORDS="0,81,101,
96">
   <AREA  SHAPE=RECT  HREF="tour.html"  COORDS="0,100,
121,117">
   <AREA  SHAPE=RECT  HREF="order.html"  COORDS="0,122,
143,138">
   <AREA  SHAPE=RECT  HREF="beginning.html"  COORDS="0,
140,213,157">

</MAP>

<CENTER>
```

```
<TABLE    BORDER="0"    CELLPADDING="0"    CELLSPACING=
"0" WIDTH="520">

<TR>

  <TD VALIGN=TOP ROWSPAN="5">
  <IMG    SRC="images/spacer.gif"    WIDTH="30"HEIGHT=
"1" BORDER="0">
  </TD>

  <TD VALIGN=TOP COLSPAN="2">
  <IMG   SRC="images/main_banner.gif"   ALT="Cryptonom
incon by Neal Stephenson" WIDTH="479" HEIGHT="122"
BORDER="0">
  </TD>

</TR>
```

This crud is called HTML (HyperText Markup Language) and it is basically a very simple programming language instructing your web browser how to draw a page on a screen. Anyone can learn HTML and many people do. The important thing is that no matter what splendid multimedia web pages they might represent, HTML files are just telegrams.

When Ronald Reagan was a radio announcer, he used to call baseball games that he did not physically attend by reading the terse descriptions that trickled in over the telegraph wire and were printed out on a paper tape. He would sit there, all by himself in a padded room with a microphone, and the paper tape would creep out of the machine and crawl over the palm of his hand printed with cryptic abbreviations. If the count went to three and two, Reagan would describe the scene as he saw it in his mind's eye: "The brawny left-hander steps out of the batter's box to wipe the sweat from his brow. The umpire steps forward to sweep the dirt from home plate," and so on. When the cryptogram on the paper tape announced a base hit, he would whack the edge of the table with a pencil, creating a little sound effect, and describe the arc of the ball as if he could actually see it. His listeners, many of whom presumably thought that Reagan was actually at the ballpark watching the game, would reconstruct the scene in their minds according to his descriptions.

This is exactly how the World Wide Web works: the HTML files are the pithy description on the paper tape, and your web browser is Ronald Reagan. The same is true of graphical user interfaces in general.

So an OS is a stack of metaphors and abstractions that stands between you and the telegrams, and embodying various tricks the programmer used to convert the information you're working with—be it images, e-mail messages, movies, or word-processing documents—into the necklaces of bytes that are the only things computers know how to work with. When we used actual telegraph equipment (teletypes) or their higher-tech substitutes ("glass teletypes," or the MS-DOS command line) to work with our computers, we were very close to the bottom of that stack. When we use most modern operating systems, though, our interaction with the machine is heavily mediated. Everything we do is interpreted and translated time and again as it works its way down through all of the metaphors and abstractions.

The Macintosh OS was a revolution in both the good and bad senses of that word. Obviously it was true that command line interfaces were not for everyone, and that it would be a good thing to make computers more accessible to a less technical audience—if not for altruistic reasons, then because those sorts of people constituted an incomparably vaster market. It was clear that the Mac's engineers saw a whole new country stretching out before them; you could almost hear them muttering, "Wow! We don't have to be bound by files as linear streams of bytes anymore, *vive la revolution*, let's see how far we can take this!" No command line interface was available on the Macintosh; you talked to it with the mouse, or not at all. This was a statement of sorts, a credential of revolutionary purity. It seemed that the designers of the Mac intended to sweep command line interfaces into the dustbin of history.

My own personal love affair with the Macintosh began in the spring of 1984 in a computer store in Cedar Rapids, Iowa, when a friend of mine—coincidentally, the son of the MGB owner—showed me a Macintosh running MacPaint, the revolutionary drawing program. It ended in July of 1995 when I tried to save a big important file on my Macintosh PowerBook and instead of doing so, it annihilated the data so thoroughly that two different disk crash utility programs were unable to find any trace that it had ever existed. During the intervening ten years, I had a passion for the MacOS that seemed righteous and reasonable at the time but in retrospect strikes me as being exactly the same sort of goofy infatuation that my friend's dad had with his car.

The introduction of the Mac triggered a sort of holy war in the computer world. Were GUIs a brilliant design innovation that made computers more human-centered and therefore accessible to the masses, leading us toward an unprecedented revolution in human society, or an insulting bit of audiovisual gimcrackery dreamed up by flaky Bay Area hacker types that stripped computers of their power and flexibility and turned the noble and serious work of computing into a childish video game?

This debate actually seems more interesting to me today than it did in the mid-1980s. But people more or less stopped debating it when Microsoft endorsed the idea of GUIs by coming out with the first Windows system. At this point, command-line partisans were relegated to the status of silly old grouches, and a new conflict was touched off: between users of MacOS and users of Windows.*

There was plenty to argue about. The first Macin-

---

* According to a rigorous, and arguably somewhat old-fashioned, definition of "operating system," Windows 95 and 98 are not operating systems at all, but rather a set of applications that run on MS-DOS, which *is* an operating system. In practice, Windows 95 and 98 are marketed and thought of as OSes and so I will tend to refer to them as such. This nomenclature is technically questionable, politically fraught, and now legally encumbered, but it is best for purposes of this essay, which is chiefly about aesthetic and cultural concerns.

toshes looked different from other PCs even when they were turned off: they consisted of one box containing both CPU (the part of the computer that does arithmetic on bits) and monitor screen. This was billed, at the time, as a philosophical statement of sorts: Apple wanted to make the personal computer into an appliance, like a toaster. But it also reflected the purely technical demands of running a graphical user interface. In a GUI machine, the chips that draw things on the screen have to be integrated with the computer's central processing unit, or CPU, to a far greater extent than is the case with command line interfaces, which until recently didn't even know that they weren't just talking to teletypes.

This distinction was of a technical and abstract nature, but it became clearer when the machine crashed. (It is commonly the case with technologies that you can get the best insight about how they work by watching them fail.) When everything went to hell and the CPU began spewing out random bits, the result, on a CLI machine, was lines and lines of perfectly formed but random characters on the screen—known to cognoscenti as "going Cyrillic." But to the MacOS, the screen was not a teletype but a place to put graphics; the image on the screen was a bitmap, a literal rendering of the contents of a particular portion of the computer's memory. When the computer crashed and wrote gibberish into the bitmap, the result was something that looked vaguely like static on a broken television set—a "snow crash."

And even after the introduction of Windows, the underlying differences endured; when a Windows machine got into trouble, the old command line interface would fall down over the GUI like an asbestos fire curtain sealing off the proscenium of a burning opera. When a Macintosh got into trouble, it presented you with a cartoon of a bomb, which was funny the first time you saw it.

These were by no means superficial differences. The reversion of Windows to a CLI when it was in distress proved to Mac partisans that Windows was nothing more than a cheap facade, like a garish afghan flung over a rotted-out sofa. They were disturbed and annoyed by the sense that lurking underneath Windows' ostensibly user-friendly interface was—literally—a subtext.

For their part, Windows fans might have made the sour observation that all computers, even Macintoshes, were built on that same subtext, and that the refusal of Mac owners to admit that fact to themselves seemed to signal a willingness, almost an eagerness, to be duped.

Anyway, a Macintosh had to switch individual bits in the memory chips on the video card, and it had to do it very fast and in arbitrarily complicated patterns. Nowadays this is cheap and easy, but in the technological regime that prevailed in the early 1980s, the only realistic way to do it was to build the motherboard (which contained the CPU) and the video system (which contained the memory that was mapped onto the screen) as a tightly integrated whole—hence the single, hermetically sealed case that made the Macintosh so distinctive.

When Windows came out, it was conspicuous for its ugliness, and its current successors, Windows 95, 98, and Windows NT, are not things that people would pay money to look at either. Microsoft's complete disregard for aesthetics gave all of us Mac-lovers plenty of opportunities to look down our noses at them. That Windows looked an awful lot like a direct ripoff of MacOS gave us a burning sense of moral outrage to go with it. Among people who really knew and appreciated computers (hackers, in Steven Levy's nonpejorative sense of that word), and in a few other niches such as professional musicians, graphic artists, and schoolteachers, the Macintosh, for a while, was simply the computer. It was seen as not only a superb piece of engineering, but an embodiment of certain ideals about the use of technology to benefit mankind, while Windows was seen as both a pathetically clumsy imitation and a sinister world domination plot rolled into one. So, very early, a pattern had been established that endures to this day: people dislike Microsoft, which is okay; but they dislike it for reasons that are poorly considered, and in the end, self-defeating.

CLASS STRUGGLE ON
THE DESKTOP

Now that the Third Rail has been firmly grasped, it is worth reviewing some basic facts here. Like any other publicly traded, for-profit corporation, Microsoft has, in effect, borrowed a bunch of money from some people (its stockholders) in order to be in the bit business. As an officer of that corporation, Bill Gates has only one responsibility, which is to maximize return on investment. He has done this incredibly well. Any actions taken in the world by Microsoft—any software released by them, for example—are basically epiphenomena, which can't be interpreted or understood except insofar as they reflect Bill Gates's execution of his one and only responsibility.

It follows that if Microsoft sells goods that are aesthetically unappealing, or that don't work very well, it does not mean that they are (respectively) philistines or halfwits. It is because Microsoft's excellent management has figured out that they can make more money for their stockholders by releasing stuff with obvious, known imperfections than they can by making it beautiful or bug-free. This is annoying, but (in the end) not half so annoying as watching Apple inscrutably and relentlessly destroy itself.

Hostility toward Microsoft is not difficult to find on the Net, and it blends two strains: resentful people who feel Microsoft is too powerful, and disdainful people who think it's tacky. This is all strongly reminiscent of the heyday of Communism and Socialism, when the bourgeoisie were hated from both ends: by the proles, because they had all the money, and by the intelligentsia, because of their tendency to spend it on lawn ornaments. Microsoft is the very embodiment of modern high-tech prosperity—it is, in a word, bourgeois—and so it attracts all of the same gripes.

The opening "splash screen" for Microsoft Word 6.0 summed it up pretty neatly: when you started up the program you were treated to a picture of an expensive enamel pen lying across a couple of sheets of fancy-looking handmade writing paper. It was obviously a bid to make the software look classy, and it might have worked for some, but it failed for me, because the pen was a ballpoint, and I'm a fountain pen man. If Apple had done it, they would've used a Mont Blanc fountain pen, or maybe a Chinese calligraphy brush. And I doubt that this was an accident. Recently I spent a while reinstalling Windows NT on one of my home computers, and many times had to double-click on the "Control Panel" icon. For reasons that are difficult to fathom, this icon consists of a picture of a clawhammer and a chisel or screwdriver resting on top of a file folder.

These aesthetic gaffes give one an almost uncontrollable urge to make fun of Microsoft, but again, it is all beside the point—if Microsoft had done focus group testing of possible alternative graphics, they probably would have found that the average mid-level office worker associated fountain pens with effete upper management toffs and was more comfortable with ballpoints. Likewise, the regular guys, the balding dads of the world who probably bear the brunt of setting up and maintaining home computers, can probably relate best to a picture of a clawhammer—while perhaps harboring fantasies of taking a real one to their balky computers.

This is the only way I can explain certain peculiar facts about the current market for operating systems, such as that ninety percent of all customers continue to buy station wagons off the Microsoft lot while free tanks are there for the taking, right across the street.

A string of ones and zeroes was not a difficult thing for Bill Gates to distribute, once he'd thought of the idea. The hard part was selling it—reassuring customers that they were actually getting something in return for their money.

Anyone who has ever bought a piece of software in a store has had the curiously deflating experience of taking the bright shrink-wrapped box home, tearing it open, finding that it's ninety-five percent air, throwing away all the little cards, party favors, and bits of trash, and loading the disk into the computer. The end result (after you've lost the disk) is nothing except some images on a computer screen, and some capabilities that weren't there before. Sometimes you don't even have that—you have a string of error messages instead. But your money is definitely gone. Now we are almost accustomed to this, but twenty years ago it was a very dicey business proposition.

Bill Gates made it work anyway. He didn't make it work by selling the best software or offering the cheapest price. Instead he somehow got people to believe that they were receiving something valuable in exchange for their money. The streets of every city in the world are filled with those hulking, rattling station wagons. Anyone who doesn't own one feels a little weird, and wonders, in spite of himself, whether it might not be time to cease resistance and buy one; anyone who does, feels confident that he has acquired some meaningful possession, even on those days when the vehicle is up on a lift in a repair shop.

All of this is perfectly congruent with membership in the bourgeoisie, which is as much a mental as a material state. And it explains why Microsoft is regularly attacked, on the Net and elsewhere, from both sides. People who are inclined to feel poor and oppressed construe everything Microsoft does as some sinister Orwellian plot. People who like to think of themselves as intelligent and informed technology users are driven crazy by the clunkiness of Windows.

Nothing is more annoying to sophisticated people than

to see someone who is rich enough to know better being tacky—unless it is to realize, a moment later, that they probably know they are tacky and they simply don't care and they are going to go on being tacky, and rich, and happy, forever. Microsoft therefore bears the same relationship to the Silicon Valley elite as the Beverly Hillbillies did to their fussy banker, Mr. Drysdale—who is irritated not so much by the fact that the Clampetts moved to his neighborhood as by the knowledge that when Jethro is seventy years old, he's still going to be talking like a hillbilly and wearing bib overalls, and he's still going to be a lot richer than Mr. Drysdale.

Even the hardware that Windows ran on, when compared to the machines put out by Apple, looked like white-trash stuff, and still mostly does. The reason was that Apple was and is a hardware company, while Microsoft was and is a software company. Apple therefore had a monopoly on hardware that could run MacOS, whereas Windows-compatible hardware came out of a free market. The free market seems to have decided that people will not pay for cool-looking computers; PC hardware makers who hire designers to make their stuff look distinctive get their clocks cleaned by Taiwanese clone makers punching out boxes that look as if they belong on cinderblocks in front of someone's trailer. Apple, on the other hand, could make their hardware as pretty as they wanted to and simply pass the higher prices on to their besotted consumers, like me. Only last week (I am writing this sentence in early January 1999) the technology sections of all the newspapers were filled with adulatory press coverage of how Apple had released the iMac in several happenin' new colors like Blueberry and Tangerine.

Apple has always insisted on having a hardware monopoly, except for a brief period in the mid-1990s when they allowed clone-makers to compete with them, before subsequently putting them out of business. Macintosh hardware was, consequently, expensive. You didn't open it up and fool around with it because doing so would void the warranty. In fact, the first Mac was specifically designed to be difficult to open—you needed a kit of exotic tools, which you could buy through little ads that began to appear in the back pages of magazines a few months after the Mac came out on the market. These ads always had a certain disreputable air about them, like pitches for lock-picking tools in the backs of lurid detective magazines.

This monopolistic policy can be explained in at least three different ways.

The charitable explanation is that the hardware monopoly policy reflected a drive on Apple's part to provide a seamless, unified blending of hardware, operating system, and software. There is something to this. It is hard enough to make an OS that works well on one specific piece of hardware, designed and tested by engineers who work down the hallway from you, in the same company. Making an OS to work on arbitrary pieces of hardware, cranked out by rabidly entrepreneurial clonemakers on the other side of the international date line, is very difficult and accounts for much of the troubles people have using Windows.

The financial explanation is that Apple, unlike Microsoft, is and always has been a hardware company. It simply depends on revenue from selling hardware, and cannot exist without it.

The not-so-charitable explanation has to do with Apple's corporate culture, which is rooted in Bay Area Baby Boomdom.

Now, since I'm going to talk for a moment about culture, full disclosure is probably in order, to protect myself against allegations of conflict of interest and ethical turpitude: (1) Geographically I am a Seattleite, of a Saturnine temperament, and inclined to take a sour view of the Dionysian Bay Area, just as they tend to be annoyed and appalled by us. (2) Chronologically I am post–Baby Boom. I feel that way, at least, because I never experienced the fun and exciting parts of the whole Boomer scene—just spent a lot of time dutifully chuckling at Boomers' maddeningly pointless anecdotes about just how stoned they got on various occasions, and politely fielding their assertions about how great their music was. But even from this remove it was possible to glean certain patterns. One that recurred as regularly as an urban legend was about how someone would move into a commune populated by sandal-wearing, peace-sign-flashing flower children and eventually discover that, underneath this facade, the guys who ran it were actually control freaks; and that, as living in a commune, where much lip service was paid to ideals of peace, love, and harmony had deprived them of normal, socially approved outlets for their control-freakdom, it tended to come out in other, invariably more sinister, ways.

Applying this to the case of Apple Computer will be left as an exercise for the reader, and not a very difficult exercise.

It is a bit unsettling, at first, to think of Apple as a control freak, because it is completely at odds with their corporate image. Weren't these the guys who aired the famous Super Bowl ads showing suited, blindfolded executives marching like lemmings off a cliff? Isn't this the company that even now runs ads picturing the Dalai Lama (except in Hong Kong) and Einstein and other offbeat rebels?

It is indeed the same company, and the fact that they have been able to plant this image of themselves as creative and rebellious freethinkers in the minds of so many intelligent and media-hardened skeptics really gives one pause. It is testimony to the insidious power of expensive slick ad campaigns and, perhaps, to a certain amount of wishful thinking in the minds of people who fall for them. It also raises the question of why Microsoft is so bad at PR, when the history of Apple demonstrates that by writing large checks to good ad agencies, you can plant a corporate image in the minds of intelligent people that is completely at odds with reality. (The answer, for people who don't like Damoclean questions, is that since Microsoft has won the hearts and minds of the silent majority—the bourgeoisie—they don't give a damn about having a slick image, any more then Dick Nixon did. "I want to believe"—the mantra that Fox Mulder has pinned to his office wall in *The X-Files*—applies in different ways to these two companies: Mac partisans want to believe in the image of Apple purveyed in those ads, and in the notion that Macs are somehow fundamentally different from other computers, while Windows people want to believe that they are getting something for their money, engaging in a respectable business transaction.)

In any event, as of 1987, both MacOS and Windows were out on the market, running on hardware platforms that were radically different from each other, not only in the sense that MacOS used Motorola CPU chips while Windows used Intel, but in the sense—then overlooked, but in the long run, vastly more significant—that the Apple hardware business was a rigid monopoly and the Windows side was a churning free-for-all.

But the full ramifications of this did not become clear until very recently—in fact, they are still unfolding, in remarkably strange ways, as I'll explain when we get to Linux. The upshot is that millions of people got accustomed to using GUIs in one form or another. By doing so, they made Apple/Microsoft a lot of money. The fortunes of many people have become bound up with the ability of these companies to continue selling products whose salability is very much open to question.

HONEY-POT, TAR-PIT, WHATEVER

When Gates and Allen invented the idea of selling software, they ran into criticism from both hackers and sober-sided businesspeople. Hackers understood that software was just information and objected to the idea of selling it. These objections were partly moral. The hackers were coming out of the scientific and academic world, where it is imperative to make the results of one's work freely available to the public. They were also partly practical: how can you sell something that can be easily copied? Businesspeople, who are polar opposites of hackers in so many ways, had objections of their own. Accustomed to selling toasters and insurance policies, they naturally had a difficult time understanding how a long collection of ones and zeroes could constitute a salable product.

Obviously Microsoft prevailed over these objections, and so did Apple. But the objections still exist. The most hackerish of all the hackers, the Ur-hacker, as it were, was and is Richard Stallman, who became so annoyed with the evil practice of selling software that in 1984 (the same year that the Macintosh went on sale) he went off and founded something called the Free Software Foundation, which commenced work on something called GNU. GNU is an acronym for Gnu's Not Unix, but this is a joke in more ways than one, because GNU most certainly *is* a functional replacement for Unix. Because of copyright concerns ("Unix" is trademarked, and the programs it comprises are copyrighted, by AT&T) they simply could not claim that it was Unix, and so, just to be extra safe, they asserted that it wasn't. Notwithstanding the incomparable talent and drive possessed by Mr. Stallman and other GNU adherents, their project to build a free Unix was a little bit like trying to dig a subway system with a teaspoon. Until, that is, the advent of Linux.*

But the basic idea of recreating an operating system from scratch was perfectly sound and completely doable. It has been done many times. It is inherent in the very nature of operating systems.

Operating systems are not strictly necessary. There is no reason why a sufficiently dedicated coder could not start from nothing with every project and write fresh code to handle such basic, low-level operations as con-

---

* Stallman insists that this OS should always be referred to as GNU/Linux, and has perfectly good reasons for saying so, viz., so that the role of the GNU project will not go unrecognized. In practice, almost everyone refers to it as Linux. For purposes of this essay I will emphasize the role of GNU by explicitly describing it rather than by using the GNU/Linux nomenclature.

trolling the read/write heads on the disk drives and lighting up pixels on the screen. The very first computers had to be programmed in this way. But since nearly every program needs to carry out those same basic operations, this approach would lead to vast duplication of effort.

Nothing is more disagreeable to the hacker than duplication of effort. The first and most important mental habit that people develop when they learn how to write computer programs is to generalize, generalize, generalize. To make their code as modular and flexible as possible, breaking large problems down into small subroutines that can be used over and over again in different contexts. Consequently, the development of operating systems, despite being technically unnecessary, was inevitable. Because at its heart, an operating system is nothing more than a library containing the most commonly used code, written once (and hopefully written well), and then made available to every coder who needs it.

So a proprietary, closed, secret operating system is a contradiction in terms. It goes against the whole point of having an operating system. And it is impossible to keep them secret anyway. The source code—the original lines of text written by the programmers—can be kept secret. But an OS as a whole is a collection of small subroutines that do very specific, very clearly defined jobs. Exactly what those subroutines do has to be made public, quite explicitly and exactly, or else the OS is completely useless to programmers; they can't make use of those subroutines if they don't have a complete and perfect understanding of what the subroutines do.

In the Beginning . . .

The only thing that isn't made public is exactly how the subroutines do what they do. But once you know what a subroutine does, it's generally quite easy (if you are a hacker) to write one of your own that does exactly the same thing. It might take a while, and it is tedious and unrewarding, but in most cases it's not really hard.

What's hard, in hacking as in fiction, is not the writing; it's deciding what to write. And the vendors of commercial OSes have already decided, and published their decisions.

This has been generally understood for a long time. MS-DOS was duplicated, functionally, by a rival product, written from scratch, called ProDOS, that did all of the same things in pretty much the same way. In other words, another company was able to write code that did all of the same things as MS-DOS and sell it at a profit. If you are using the Linux OS, you can get a free program called WINE, which is a Windows emulator; that is, you can open up a window on your desktop that runs Windows programs. It means that a completely functional Windows OS has been recreated inside of Unix, like a ship in a bottle. And Unix itself, which is vastly more sophisticated than MS-DOS, has been built up from scratch many times over. Versions of it are sold by Sun, Hewlett-Packard, AT&T, Silicon Graphics, IBM, and others.

People have, in other words, been rewriting basic OS code for so long that all of the technology that constituted an "operating system" in the traditional (pre-GUI) sense of that phrase is now so cheap and common that it's literally free. Not only could Gates and Allen not sell MS-DOS today, they could not even give it away, because much more powerful OSes are already being given away. Even the original Windows has become worthless, in that there is no point in owning something that can be emulated inside of Linux—which is, itself, free.

In this way the OS business is very different from, say, the car business. Even an old rundown car has some value. You can use it for making runs to the dump, or strip it for parts. It is the fate of manufactured goods to slowly and gently depreciate as they get old and have to compete against more modern products.

But it is the fate of operating systems to become free.

Microsoft is a great software applications company. Applications—such as Microsoft Word—are an area where innovation brings real, direct, tangible benefits to users. The innovations might be new technology straight from the research department, or they might be in the category of bells and whistles, but in any event they are frequently useful and they seem to make users happy. And Microsoft is in the process of becoming a great research company. But Microsoft is not such a great operating systems company. This is not necessarily because their operating systems are all that bad from a purely technological standpoint. Microsoft's OSes do have their problems, sure, but they are vastly better than they used to be, and they are adequate for most people.

Why, then, do I say that Microsoft is not such a great operating systems company? Because the very nature of operating systems is such that it is senseless for them to be developed and owned by a specific company. It's a thankless job to begin with. Applications create possibilities for millions of credulous users, whereas OSes impose limitations on thousands of grumpy coders, and so OS-makers will forever be on the shit-list of anyone who counts for anything in the high-tech world. Applications get used by people whose big problem is understanding all of their features, whereas OSes get hacked by coders who are annoyed by their limitations. The OS business has been good to Microsoft only insofar as it has given them the money they needed to launch a really good applications software business and to hire a lot of smart researchers. Now it really ought to be jettisoned, like a spent booster stage from a rocket. The big question is whether Microsoft is capable of doing this. Or is it addicted to OS sales in the same way as Apple is to selling hardware?

Keep in mind that Apple's ability to monopolize its own hardware supply was once cited, by learned observers, as a great advantage over Microsoft. At the time, it seemed to place them in a much stronger position. In the end, it nearly killed them, and may kill them yet. The problem, for Apple, was that most of the world's computer users ended up owning cheaper hardware. But cheap hardware couldn't run MacOS, and so these people switched to Windows.

Replace "hardware" with "operating systems," and "Apple" with "Microsoft," and you can see the same thing about to happen all over again. Microsoft dominates the OS market, which makes them money and seems like a great idea for now. But cheaper and better OSes are available, and they are growingly popular in parts of the world that are not so saturated with computers as the U.S. Ten years from now, most of the world's computer users may end up owning these cheaper OSes. But these OSes do not, for the time being, run any Microsoft applications, and so these people will use something else.

To put it more directly: every time someone decides to use a non-Microsoft OS, Microsoft's OS division, obviously, loses a customer. But, as things stand now, Microsoft's applications division loses a customer too. This is not such a big deal as long as almost everyone uses Microsoft OSes. But as soon as Windows' market share begins to slip, the math starts to look pretty dismal for the people in Redmond.

This argument could be countered by saying that Microsoft could simply recompile its applications to run under other OSes. But this strategy goes against most normal corporate instincts. Again the case of Apple is instructive. When things started to go south for Apple, they should have ported their OS to cheap PC hardware. But they didn't. Instead, they tried to make the most of their brilliant hardware, adding new features and expanding the product line. But this only had the effect of making their OS more dependent on these special hardware features, which made it worse for them in the end.

Likewise, when Microsoft's position in the OS world is threatened, their corporate instincts will tell them to

pile more new features into their operating systems, and
then re-jigger their software applications to exploit those
special features. But this will only have the effect of
making their applications dependent on an OS with de-
clining market share, and make it worse for them in
the end.

The operating system market is a death trap, a tar pit, a
slough of despond. There are only two reasons to invest
in Apple and Microsoft. (1) Each of these companies is in
what we would call a codependency relationship with
their customers. The customers Want To Believe, and
Apple and Microsoft know how to give them what they
want. (2) Each company works very hard to add new
features to their OSes, which works to secure customer
loyalty, at least for a little while.

Accordingly, most of the remainder of this essay will
be about those two topics.

THE TECHNOSPHERE

Unix is the only OS remaining whose GUI (a vast suite of code called the XWindow System) is separate from the OS in the old sense of the phrase. This is to say that you can run Unix in pure command line mode if you want to, with no windows, icons, mouses, etc. whatsoever, and it will still be Unix and capable of doing everything Unix is supposed to do. But the other OSes—MacOS, the Windows family, and BeOS—have their GUIs tangled up with the old-fashioned OS functions to the extent that they have to run in GUI mode, or else they are not really running. So it's no longer really possible to think of GUIs as being distinct from the OS; they're now an inextricable part of the OSes that they belong to—and they are by far the largest part, and by far the most expensive and difficult part to create.

There are only two ways to sell a product: price and features. When OSes are free, OS companies cannot compete on price, and so they compete on features. This means that they are always trying to outdo each other writing code that, until recently, was not considered to be part of an OS at all: stuff like GUIs. This explains a lot about how these companies behave.

It explains why Microsoft added a browser to their OS, for example. It is easy to get free browsers, just as to get free OSes. If browsers are free, and OSes are free, it would seem that there is no way to make money from browsers or OSes. But if you can integrate a browser into the OS and thereby imbue both of them with new features, you have a salable product.

Setting aside, for the moment, the fact that this makes government antitrust lawyers really mad, this strategy makes sense. At least, it makes sense if you assume (as Microsoft's management appears to) that the OS has to be protected at all costs. The real question is whether every new technological trend that comes down the pike ought to be used as a crutch to maintain the OS's dominant position. Confronted with the web phenomenon, Microsoft had to develop a really good web browser, and they did. But then they had a choice: they could have made that browser work on many different OSes, which would give Microsoft a strong position in the Internet world no matter what happened to their OS market share. Or they could make the browser appear to be one with the OS, gambling that this would make the OS look so modern and sexy that it would help to preserve their dominance in that market. The problem is that when Microsoft's OS position begins to erode (and since it is currently at something like ninety percent, it can't go anywhere but down) it will drag everything else down with it.

In your high school geology class you probably were taught that all life on earth exists in a paper-thin shell called the biosphere, which is trapped between thousands of miles of dead rock underfoot, and cold dead radioactive empty space above. Companies that sell OSes exist in a sort of technosphere. Underneath is technology that has already become free. Above is technology that has yet to be developed, or that is too crazy and speculative to be productized just yet. Like the earth's biosphere, the technosphere is very thin compared to what is above and what is below.

But it moves a lot faster. In various parts of our world, it is possible to go and visit rich fossil beds where skeleton lies piled upon skeleton, recent ones on top and more ancient ones below. In theory they go all the way back to the first single-celled organisms. And if you use your imagination a bit, you can understand that, if you hang around long enough, you'll become fossilized there too, and in time some more advanced organism will become fossilized on top of you.

The fossil record—the La Brea Tar Pit—of software technology is the Internet. Anything that shows up there is free for the taking (possibly illegal, but free). Executives at companies like Microsoft must get used to the experience—unthinkable in other industries—of throwing millions of dollars into the development of new technologies, such as web browsers, and then seeing the same or equivalent software show up on the Internet for free two years, or a year, or even just a few months, later.

By continuing to develop new technologies and add

features onto their products, they can keep one step ahead of the fossilization process, but on certain days they must feel like mammoths caught at La Brea, using all their energies to pull their feet, over and over again, out of the sucking hot tar that wants to cover and envelop them.

Survival in this biosphere demands sharp tusks and heavy, stomping feet at one end of the organization, and Microsoft famously has those. But trampling the other mammoths into the tar can only keep you alive for so long. The danger is that in their obsession with staying out of the fossil beds, these companies will forget about what lies above: the realm of new technology. In other words, they must hang on to their primitive weapons and crude competitive instincts, but also evolve powerful brains. This appears to be what Microsoft is doing with its research division, which has been hiring smart people right and left. (Here I should mention that although I know, and socialize with, several people in that company's research division, we never talk about business issues and I have little to no idea what the hell they are up to. I have learned much more about Microsoft by using the Linux operating system than I ever would have done by using Windows.)

Never mind how Microsoft used to make money; today, it is making its money on a kind of temporal arbitrage. "Arbitrage," in the usual sense, means to make money by taking advantage of differences in the price of something between different markets. It is spatial, in other words, and hinges on the arbitrageur knowing what is going on simultaneously in different places. Microsoft is making money by taking advantage of differences in the price of technology in different times. Temporal arbitrage, if I may coin a phrase, hinges on the arbitrageur knowing what technologies people will pay money for next year, and how soon afterwards those same technologies will become free. What spatial and temporal arbitrage have in common is that both hinge on the arbitrageur's being extremely well informed: one about price gradients across space at a given time, and the other about price gradients over time in a given place.

So Apple and Microsoft shower new features upon their users almost daily, in the hopes that a steady stream of genuine technical innovations, combined with the "I want to believe" phenomenon, will prevent their customers from looking across the road toward the cheaper and better OSes that are available to them. The question is whether this makes sense in the long run. If Microsoft is addicted to OSes as Apple is to hardware, then they will bet the whole farm on their OSes and tie all of their new applications and technologies to them. Their continued survival will then depend on these two things: adding more features to their OSes so that customers will not switch to the cheaper alternatives, and maintaining the image that, in some mysterious way, gives those customers the feeling that they are getting something for their money.

The latter is a truly strange and interesting cultural phenomenon.

THE INTERFACE
CULTURE*

A few years ago I walked into a grocery store some-where and was presented with the following *tableau vivant*: near the entrance a young couple were standing in front of a large cosmetics display. The man was stolidly holding a shopping basket between his hands while his mate raked blister-packs of makeup off the display and piled them in. Since then I've always thought of that man as the personification of an interesting human tendency: not only are we not offended to be dazzled by manufactured images, but we like it. We practically insist on it. We are eager to be complicit in our own dazzlement: to pay money for a theme park ride, vote for a guy who's obviously lying to us, or stand there holding the basket as it's filled up with cosmetics.

I was in Disney World recently, specifically the part

---

* Apologies for this section title to Steven Johnson, author of *Interface Culture: How New Technology Transforms the Way We Create and Communicate,* Harper San Francisco (1997) and Basic Books (1999).

of it called the Magic Kingdom, walking up Main Street USA. This is a perfect gingerbready Victorian small town that culminates in a Disney castle. It was very crowded; we shuffled rather than walked. Directly in front of me was a man with a camcorder. It was one of the new breed of camcorders where instead of peering through a viewfinder you gaze at a flat-panel color screen about the size of a playing card, which televises live coverage of whatever the camcorder is seeing. He was holding the appliance close to his face, so that it obstructed his view. Rather than go see a real small town for free, he had paid money to see a pretend one, and rather than see it with the naked eye, he was watching it on television.

And rather than stay home and read a book, I was watching him.

Americans' preference for mediated experiences is obvious enough, and I'm not going to keep pounding it into the ground. I'm not even going to make snotty comments about it—after all, I was at Disney World as a paying customer. But it clearly relates to the colossal success of GUIs, and so I have to talk about it some. Disney does mediated experiences better than anyone. If they understood what OSes are, and why people use them, they could crush Microsoft in a year or two.

In the part of Disney World called the Animal Kingdom there is a new attraction called the Maharajah Jungle Trek. It was open for sneak previews when I was there. This is a complete stone-by-stone reproduction of a hypothetical ruin in the jungles of India. According to its backstory, it was built by a local rajah in the sixteenth century as a game reserve. He would go there with his princely guests to hunt Bengal tigers. As time went on, it fell into disrepair and the tigers and monkeys took it over; eventually, around the time of India's independence, it became a government wildlife reserve, now open to visitors.

The place looks more like what I have just described than any actual building you might find in India. All the stones in the broken walls are weathered as if monsoon rains had been trickling down them for centuries, the paint on the gorgeous murals is flaked and faded just so, and Bengal tigers loll amid stumps of broken columns. Where modern repairs have been made to the ancient structure, they've been done, not as Disney's engineers would do them, but as thrifty Indian janitors would— with hunks of bamboo and rust-spotted hunks of rebar. The rust is painted on, of course, and protected from real rust by a plastic clear-coat, but you can't tell unless you get down on your knees.

In one place you walk along a stone wall with a series of old pitted friezes carved into it. One end of the wall has broken off and settled into the earth, perhaps because of some long-forgotten earthquake, and so a broad jagged crack runs across a panel or two, but the story is still readable: first, primordial chaos leads to a flourishing of many animal species. Next, we see the Tree of Life surrounded by diverse animals. This is an obvious allusion (or, in showbiz lingo, a tie-in) to the gigantic Tree of Life that dominates the center of Disney's Animal

Kingdom just as the Castle dominates the Magic Kingdom or the Sphere does Epcot. But it's rendered in historically correct style and could probably fool anyone who didn't have a Ph.D. in Indian art history.

The next panel shows a mustachioed H. sapiens chopping down the Tree of Life with a scimitar, and the animals fleeing every which way. The one after that shows the misguided human getting walloped by a tidal wave, part of a latter-day Deluge presumably brought on by his stupidity.

The final panel, then, portrays the Sapling of Life beginning to grow back, but now Man has ditched the edged weapon and joined the other animals in standing around to adore and praise it.

It is, in other words, a prophecy of the Bottleneck: the scenario, commonly espoused among modern-day environmentalists, that the world faces an upcoming period of grave ecological tribulations that will last for a few decades or centuries and end when we find a new harmonious modus vivendi with Nature.

Taken as a whole the frieze is a pretty brilliant piece of work. Obviously it's not an ancient Indian ruin, and some person or people now living deserve credit for it. But there are no signatures on the Maharajah's game reserve at Disney World. There are no signatures on anything, because it would ruin the whole effect to have long strings of production credits dangling from every custom-worn brick, as they do from Hollywood movies.

Among Hollywood writers, Disney has the reputation of being a real wicked stepmother. It's not hard to see why. Disney is in the business of putting out a product of seamless illusion—a magic mirror that reflects the world back better than it really is. But a writer is literally talking to his or her readers, not just creating an ambience or presenting them with something to look at. Just as the command line interface opens a much more direct and explicit channel from user to machine than the GUI, so it is with words, writer, and reader.

The word, in the end, is the only system of encoding thoughts—the only medium—that is not fungible, that refuses to dissolve in the devouring torrent of electronic media. (The richer tourists at Disney World wear t-shirts printed with the names of famous designers, because designs themselves can be bootlegged easily and with impunity. The only way to make clothing that cannot be legally bootlegged is to print copyrighted and trademarked words on it; once you have taken that step, the clothing itself doesn't really matter, and so a t-shirt is as good as anything else. T-shirts with expensive words on them are now the insignia of the upper class. T-shirts with cheap words, or no words at all, are for the commoners.)

But this special quality of words and of written communication would have the same effect on Disney's product as spray-painted graffiti on a magic mirror. So Disney does most of its communication without resorting to words, and for the most part, the words aren't missed. Some of Disney's older properties, such as Peter Pan, Winnie the Pooh, and Alice in Wonderland, came out of books. But the authors' names are rarely if ever mentioned, and you can't buy the original books at the Disney store. If you could, they would all seem old and queer, like very bad knockoffs of the purer, more authentic Disney versions. Compared to more recent productions like *Beauty and the Beast* and *Mulan*, the Disney movies based on these books (particularly *Alice in Wonderland* and *Peter Pan*) seem deeply bizarre, and not wholly appropriate for children. That stands to reason, because Lewis Carroll and J. M. Barrie were very strange men, and such is the nature of the written word that their personal strangeness shines straight through all the layers of Disneyfication like X-rays through a wall. Probably for this very reason, Disney seems to have stopped buying rights to books altogether, and now finds its themes and characters in folk tales, which have the lapidary, timeworn quality of the ancient bricks in the Maharajah's ruins.

If I can risk a broad generalization, most of the people who go to Disney World have zero interest in absorbing new ideas from books. This sounds snide, but listen: they have no qualms about being presented with ideas in other forms. Disney World is stuffed with environmental messages now, and the guides at Animal Kingdom can talk your ear off about biology.

If you followed those tourists home, you might find art, but it would be the sort of unsigned folk art that's for sale in Disney World's African- and Asian-themed stores. In general they only seem comfortable with media that have been ratified by great age, massive popular acceptance, or both. In this world, artists are like the anonymous, illiterate stone carvers who built the great cathedrals of Europe and then faded away into unmarked graves in the churchyard. The cathedral as a whole is awesome and stirring in spite, and possibly because, of the fact that we have no idea who built it. When we walk through it, we are communing not with individual stone carvers but with an entire culture.

Disney World works the same way. If you are an intellectual type, a reader or writer of books, the nicest thing you can say about this is that the execution is superb. But it's easy to find the whole environment a little creepy, because something is missing: the translation of all its content into clear explicit written words, the attribution of the ideas to specific people. You can't argue with it. It seems as if a hell of a lot might be being glossed over, as if Disney World might be putting one over on us, and possibly getting away with all kinds of buried assumptions and muddled thinking.

And this is precisely the same as what is lost in the transition from the command line interface to the GUI.

Disney and Apple/Microsoft are in the same business: short-circuiting laborious, explicit verbal communication with expensively designed interfaces. Disney is a sort of user interface unto itself—and more than just graphical. Let's call it a Sensorial Interface. It can be applied to anything in the world, real or imagined, albeit at staggering expense.

Why are we rejecting explicit word-based interfaces, and embracing graphical or sensorial ones—a trend that accounts for the success of both Microsoft and Disney?

Part of it is simply that the world is very complicated now—much more complicated than the hunter–gatherer world that our brains evolved to cope with—and we simply can't handle all of the details. We have to delegate. We have no choice but to trust some nameless artist at Disney or programmer at Apple or Microsoft to make a few choices for us, close off some options, and give us a conveniently packaged executive summary.

But more importantly, it comes out of the fact that during this century, intellectualism failed, and everyone knows it. In places like Russia and Germany, the common people agreed to loosen their grip on traditional folkways, mores, and religion, and let the intellectuals run with the ball, and they screwed everything up and turned the century into an abattoir. Those wordy intellectuals used to be merely tedious; now they seem kind of dangerous as well.

We Americans are the only ones who didn't get creamed at some point during all of this. We are free and prosperous because we have inherited political and value systems fabricated by a particular set of eighteenth-century intellectuals who happened to get it right. But we have lost touch with those intellectuals, and with anything like intellectualism, even to the point of not reading books anymore, though we are literate. We seem much more comfortable with propagating those values to future generations nonverbally, through a process of being steeped in media. Apparently this actually works to some degree, for police in many lands are now complaining that local arrestees are insisting on having their Miranda rights read to them, just like perps in American TV cop shows. When it's explained to them that they are in a different country, where those rights do not exist, they become outraged. *Starsky and Hutch* reruns, dubbed into diverse languages, may turn out, in the long run, to be a greater force for human rights than the Declaration of Independence.

The written word is unique among media in that it is a digital medium that humans can, nonetheless, easily read and write. Humans are conversant in many media (music, dance, painting), but all of them are analog except for the written word, which is naturally expressed in digital form (i.e. it is a series of discrete symbols—every letter in every book is a member of a certain character set, every "a" is the same as every other "a," and so on). As any communications engineer can tell you, digital signals are much better to work with than analog ones because they are easily copied, transmitted, and error-checked. Unlike analog signals, they are not doomed to degradation over time and distance. That is why digital compact disks replaced analog LPs, for example. The digital nature of the written word confers on it exceptional stability, which is why it is the vehicle of choice for extremely important concepts like the Ten Commandments, the Koran, and the Bill of Rights. This is generally thought to be a rather good idea. But the messages conveyed by modern audiovisual media cannot be pegged to any fixed, written set of precepts in that way and consequently they are free to wander all over the place and possibly dump loads of crap into people's minds.

Orlando used to have a military installation called McCoy Air Force Base, with long runways from which B-52s could take off and reach Cuba, or just about anywhere else, with loads of nukes. But now McCoy has been scrapped and re-purposed. It has been absorbed into Orlando's civilian airport. The long runways are being used to land 747-loads of tourists from Brazil, Italy, Russia, and Japan, so that they can come to Disney World and steep in our media for a while.

To traditional cultures, especially word-based ones such as Islam, this is infinitely more threatening than the B-52s ever were. It is obvious, to everyone outside of the United States, that our arch-buzzwords—multiculturalism and diversity—are false fronts that are being used (in many cases unwittingly) to conceal a global trend to eradicate cultural differences. The basic tenet of multiculturalism (or "honoring diversity" or whatever you want to call it) is that people need to stop judging each other—to stop asserting (and, eventually, to stop believing) that this is right and that is wrong, this true and that false, one thing ugly and another thing beautiful, that God exists and has this or that set of qualities.

The lesson most people are taking home from the twentieth century is that, in order for a large number of different cultures to coexist peacefully on the globe (or even in a neighborhood) it is necessary for people to suspend judgment in this way. Hence (I would argue) our suspicion of, and hostility toward, all authority figures in modern culture. As David Foster Wallace has explained in his essay "E Unibus Pluram," this is the fundamental message of television; it is the message that people absorb, anyway, after they have steeped in our media long enough. It's not expressed in these highfalutin terms, of course. It comes through as the presumption that all authority figures—teachers, generals, cops, ministers, politicians—are hypocritical buffoons, and that hip jaded coolness is the only way to be.

The problem is that once you have done away with the ability to make judgments as to right and wrong, true and false, etc., there's no real culture left. All that remains is clog dancing and macrame. The ability to make judgments, to believe things, is the entire point of having a culture. I think this is why guys with machine guns sometimes pop up in places like Luxor and begin pumping bullets into Westerners. They perfectly understand the lesson of McCoy Air Force Base. When their sons come home wearing Chicago Bulls caps with the bills turned sideways, the dads go out of their minds.

The global anticulture that has been conveyed into every cranny of the world by television is a culture unto itself, and by the standards of great and ancient cultures like Islam and France, it seems grossly inferior, at least at first. The only good thing you can say about it is that it makes world wars and Holocausts less likely—and that is actually a pretty good thing!

The only real problem is that anyone who has no culture, other than this global monoculture, is completely screwed. Anyone who grows up watching TV, never sees any religion or philosophy, is raised in an atmosphere of moral relativism, learns about civics from watching bimbo eruptions on network TV news, and attends a university where postmodernists vie to outdo each other in demolishing traditional notions of truth and quality, is going to come out into the world as one pretty feckless human being. And—again—perhaps the goal of all this is to make us feckless so we won't nuke each other.

On the other hand, if you are raised within some specific culture, you end up with a basic set of tools that you can use to think about and understand the world. You might use those tools to reject the culture you were raised in, but at least you've got some tools.

In this country, the people who run things—who populate major law firms and corporate boards—understand all of this at some level. They pay lip service to multiculturalism and diversity and nonjudgmentalness, but they don't raise their own children that way. I have highly educated, technically sophisticated friends who have moved to small towns in Iowa to live and raise their children, and there are Hasidic Jewish enclaves in New York where large numbers of kids are being brought up according to traditional beliefs. Any suburban community might be thought of as a place where people who hold certain (mostly implicit) beliefs go to live among others who think the same way.

And not only do these people feel some responsibility to their own children, but to the country as a whole. Some of the upper class are vile and cynical, of course, but many spend at least part of their time fretting about what direction the country is going in and what responsibilities they have. And so issues that are important to book-reading intellectuals, such as global environmental collapse, eventually percolate through the porous buffer of mass culture and show up as ancient Hindu ruins in Orlando.

In the Beginning . . .

You may be asking: what the hell does all this have to do with operating systems? As I've explained, there is no way to explain the domination of the OS market by Apple/Microsoft without looking to cultural explanations, and so I can't get anywhere, in this essay, without first letting you know where I'm coming from vis-à-vis contemporary culture.

Contemporary culture is a two-tiered system, like the Morlocks and the Eloi in H. G. Wells's *The Time Machine,* except that it's been turned upside down. In *The Time Machine,* the Eloi were an effete upper class, supported by lots of subterranean Morlocks who kept the technological wheels turning. But in our world it's the other way round. The Morlocks are in the minority, and they are running the show, because they understand how everything works. The much more numerous Eloi learn everything they know from being steeped from birth in electronic media directed and controlled by book-reading Morlocks. That many ignorant people could be dangerous if they got pointed in the wrong direction, and so we've evolved a popular culture that is (a) almost unbelievably infectious, and (b) neuters every person who gets infected by it, by rendering them unwilling to make judgments and incapable of taking stands.

Morlocks, who have the energy and intelligence to comprehend details, go out and master complex subjects and produce Disney-like Sensorial Interfaces so that Eloi can get the gist without having to strain their minds or endure boredom. Those Morlocks will go to India and tediously explore a hundred ruins, then come home and build sanitary bug-free versions: highlight films, as it were. This costs a lot, because Morlocks insist on good coffee and first-class airline tickets, but that's no problem, because Eloi like to be dazzled and will gladly pay for it all.

Now I realize that most of this probably sounds snide and bitter to the point of absurdity: your basic snotty intellectual throwing a tantrum about those unlettered philistines. As if I were a self-styled Moses, coming down from the mountain all alone, carrying the stone tablets bearing the Ten Commandments carved in immutable stone—the original command line interface—and blowing his stack at the weak, unenlightened Hebrews worshipping images. Not only that, but it sounds like I'm pumping some sort of conspiracy theory.

But that is not where I'm going with this. The situation I describe here could be bad, but doesn't have to be bad and isn't necessarily bad now.

It simply is the case that we are way too busy, nowadays, to comprehend everything in detail. And it's better to comprehend it dimly, through an interface, than not at all. Better for ten million Eloi to go on the Kilimanjaro Safari at Disney World than for a thousand cardiovascular surgeons and mutual fund managers to go on "real" ones in Kenya. The boundary between these two classes is more porous than I've made it sound. I'm always running into regular dudes—construction workers, auto mechanics, taxi drivers, galoots in general—who were largely aliterate until something made it necessary for them to become readers and start actually thinking about things. Perhaps they had to come to grips with alcoholism, perhaps they got sent to jail, or came down with a disease, or suffered a crisis in religious faith, or simply got bored. Such people can get up to speed on particular subjects quite rapidly. Sometimes their lack of a broad education makes them overapt to go off on intellectual wild-goose chases, but hey, at least a wild-goose chase gives you some exercise. The spectre of a polity controlled by the fads and whims of voters who actually believe that there are significant differences between Bud Lite and Miller Lite, and who think that professional wrestling is for real, is naturally alarming to people who don't. But then countries controlled via the command line interface, as it were, by double-domed intellectuals, be they religious or secular, are generally miserable places to live.

Sophisticated people deride Disneyesque entertainments as pat and saccharine, but if the result of that is to instill basically warm and sympathetic reflexes, at a preverbal level, into hundreds of millions of unlettered media-steepers, then how bad can it be? We killed a lobster in our kitchen last night and my daughter cried for an hour. The Japanese, who used to be just about the fiercest people on earth, have become infatuated with cuddly, adorable cartoon characters. My own family—the people I know best—is divided about evenly between people who will probably read this essay and people who almost certainly won't, and I can't say for sure that one group is necessarily warmer, happier, or better-adjusted than the other.

MORLOCKS AND ELOI AT
THE KEYBOARD

Back in the days of the command line interface, users were all Morlocks who had to convert their thoughts into alphanumeric symbols and type them in, a grindingly tedious process that stripped away all ambiguity, laid bare all hidden assumptions, and cruelly punished laziness and imprecision. Then the interface-makers went to work on their GUIs and introduced a new semiotic layer between people and machines. People who use such systems have abdicated the responsibility, and surrendered the power, of sending bits directly to the chip that's doing the arithmetic, and handed that responsibility and power over to the OS. This is tempting, because giving clear instructions, to anyone or anything, is difficult. We cannot do it without thinking, and depending on the complexity of the situation, we may have to think hard about abstract things, and consider any number of ramifications, in order to do a good job of it. For most of us, this is hard work. We want things to be easier. How badly we want it can be measured by the size of Bill Gates's fortune.

In the Beginning . . .

The OS has (therefore) become a sort of intellectual labor-saving device that tries to translate humans' vaguely expressed intentions into bits. In effect we are asking our computers to shoulder responsibilities that have always been considered the province of human beings—we want them to understand our desires, to anticipate our needs, to foresee consequences, to make connections, to handle routine chores without being asked, to remind us of what we ought to be reminded of while filtering out noise.

At the upper (which is to say, closer to the user) levels, this is done through a set of conventions—menus, buttons, and so on. These work in the sense that analogies work: they help Eloi understand abstract or unfamiliar concepts by likening them to something known. But the loftier word "metaphor" is used.

The overarching concept of the MacOS was the "desktop metaphor," and it subsumed any number of lesser (and frequently conflicting, or at least mixed) metaphors. Under a GUI, a file (frequently called "document") is metaphrased as a window on the screen (which is called a "desktop"). The window is almost always too small to contain the document and so you "move around," or, more pretentiously, "navigate" in the document by "clicking and dragging" the "thumb" on the "scroll bar." When you "type" (using a keyboard) or "draw" (using a "mouse") into the "window" or use pull-down "menus" and "dialog boxes" to manipulate its contents, the results of your labors get stored (at least in theory) in a "file," and later you can pull the same information back up into another "window." When you don't want it anymore, you "drag" it into the "trash."

There is massively promiscuous metaphor-mixing going on here, and I could deconstruct it till the cows come home, but I won't. Consider only one word: "document." When we document something in the real world, we make fixed, permanent, immutable records of it. But computer documents are volatile, ephemeral constellations of data. Sometimes (as when you've just opened or saved them) the document as portrayed in the window is identical to what is stored, under the same name, in a file on the disk, but other times (as when you have made changes without saving them) it is completely different. In any case, every time you hit "Save" you annihilate the previous version of the "document" and replace it with whatever happens to be in the window at the moment. So even the word "save" is being used in a sense that is grotesquely misleading—"destroy one version, save another" would be more accurate.

Anyone who uses a word processor for very long inevitably has the experience of putting hours of work into a long document and then losing it because the computer crashes or the power goes out. Until the moment that it disappears from the screen, the document seems every bit as solid and real as if it had been typed out in ink on paper. But in the next moment, without warning, it is completely and irretrievably gone, as if it had never existed. The user is left with a feeling of disorientation (to say nothing of annoyance) stemming from a kind of metaphor shear—you realize that you've been living and thinking inside of a metaphor that is essentially bogus.

So GUIs use metaphors to make computing easier, but they are bad metaphors. Learning to use them is essentially a word game, a process of learning new definitions of words such as "window" and "document" and "save" that are different from, and in many cases almost diametrically opposed to, the old. Somewhat improbably, this has worked very well, at least from a commercial standpoint, which is to say that Apple/Microsoft have made a lot of money off of it. All of the other modern operating systems have learned that in order to be accepted by users they must conceal their underlying gutwork beneath the same sort of spackle. This has some advantages: if you know how to use one GUI operating system, you can probably work out how to use any other in a few minutes. Everything works a little differently, like European plumbing—but with some fiddling around, you can type a memo or surf the web.

Most people who shop for OSes (if they bother to shop at all) are comparing not the underlying functions but the superficial look and feel. The average buyer of an OS is not really paying for, and is not especially interested in, the low-level code that allocates memory or writes bytes onto the disk. What we're really buying is a system of metaphors. And—much more important—what we're buying into is the underlying assumption that metaphors are a good way to deal with the world.

Recently a lot of new hardware has become available that gives computers numerous interesting ways of af-

fecting the real world: making paper spew out of printers, causing words to appear on screens thousands of
miles away, shooting beams of radiation through cancer
patients, creating realistic moving pictures of the *Titanic*.
Windows is now used as an OS for cash registers and
bank tellers' terminals. My satellite TV system uses a
sort of GUI to change channels and show program
guides. Modern cellular telephones have a crude GUI
built into a tiny LCD screen. Even Legos now have a
GUI: you can buy a Lego set called Mindstorms that
enables you to build little Lego robots and program
them through a GUI on your computer. So we are now
asking the GUI to do a lot more than serve as a glorified
typewriter. Now we want it to become a generalized tool
for dealing with reality. This has become a bonanza for
companies that make a living out of bringing new technology to the mass market.

Obviously you cannot sell a complicated technological
system to people without some sort of interface that enables them to use it. The internal combustion engine was
a technological marvel in its day, but useless as a consumer good until a clutch, transmission, steering wheel,
and throttle were connected to it. That odd collection of
gizmos, which survives to this day in every car on the
road, made up what we would today call a user interface.
But if cars had been invented after Macintoshes, carmakers would not have bothered to gin up all of these
arcane devices. We would have a computer screen instead of a dashboard, and a mouse (or at best a joystick)

instead of a steering wheel, and we'd shift gears by pulling down a menu:

PARK

REVERSE

NEUTRAL

3
2
1

Help . . .

A few lines of computer code can thus be made to substitute for any imaginable mechanical interface. The problem is that in many cases the substitute is a poor one. Driving a car through a GUI would be a miserable experience. Even if the GUI were perfectly bug-free, it would be incredibly dangerous, because menus and buttons simply can't be as responsive as direct mechanical controls. My friend's dad, the gentleman who was restoring the MGB, never would have bothered with it if it had been equipped with a GUI. It wouldn't have been any fun.

The steering wheel and gearshift lever were invented during an era when the most complicated technology in most homes was a butter churn. Those early carmakers were simply lucky, in that they could dream up whatever

interface was best suited to the task of driving an automobile, and people would learn it. Likewise with the dial
telephone and the AM radio. By the time of the Second
World War, most people knew several interfaces: they
could not only churn butter but also drive a car, dial a
telephone, turn on a radio, summon flame from a ciga-
rette lighter, and change a lightbulb.

But now every little thing—wristwatches, VCRs,
stoves—is jammed with features, and every feature is
useless without an interface. If you are like me, and like
most other consumers, you have never used ninety percent of the available features on your microwave oven,
VCR, or cell phone. You don't even know that these
features exist. The small benefit they might bring you is
outweighed by the sheer hassle of having to learn about
them. This has got to be a big problem for makers of
consumer goods, because they can't compete without offering features.

It's no longer acceptable for engineers to invent a
wholly novel user interface for every new product, as
they did in the case of the automobile, partly because
it's too expensive and partly because ordinary people
can only learn so much. If the VCR had been invented
a hundred years ago, it would have come with a thumb-
wheel to adjust the tracking and a gearshift to change
between forward and reverse, and a big cast-iron handle
to load or to eject the cassettes. It would have had a big
analog clock on the front of it, and you would have set
the time by moving the hands around on the dial. But
because the VCR was invented when it was—during a

sort of awkward transitional period between the era of mechanical interfaces and GUIs—it just had a bunch of pushbuttons on the front, and in order to set the time you had to push the buttons in just the right way. This must have seemed reasonable enough to the engineers responsible for it, but to many users it was simply impossible. Thus the famous blinking 12:00 that appears on so many VCRs. Computer people call this "the blinking twelve problem." When they talk about it, though, they usually aren't talking about VCRs.

Modern VCRs usually have some kind of on-screen programming, which means that you can set the time and control other features through a sort of primitive GUI. GUIs have virtual pushbuttons too, of course, but they also have other types of virtual controls, like radio buttons, checkboxes, text entry boxes, dials, and scrollbars. Interfaces made out of these components seem to be a lot easier, for many people, than pushing those little buttons on the front of the machine, and so the blinking 12:00 itself is slowly disappearing from America's living rooms. The blinking twelve *problem* has moved on to plague other technologies.

So the GUI has gone beyond being an interface to personal computers, and has become a sort of meta-interface that is pressed into service for every new piece of consumer technology. It is rarely an ideal fit, but having an ideal, or even a good, interface is no longer the priority; the important thing now is having some kind of interface that customers will actually use, so that manufacturers can claim, with a straight face, that they are offering new features.

We want GUIs largely because they are convenient and because they are easy—or at least the GUI makes it seem that way. Of course, nothing is really easy and simple, and putting a nice interface on top of it does not change that fact. A car controlled through a GUI would be easier to drive than one controlled through pedals and steering wheel, but it would be incredibly dangerous.

By using GUIs all the time we have insensibly bought into a premise that few people would have accepted if it were presented to them bluntly: namely, that hard things can bc made easy, and complicated things simple, by putting the right interface on them. In order to understand how bizarre this is, imagine that book reviews were written according to the same values system that we apply to user interfaces: "The writing in this book is marvelously simple-minded and glib; the author glosses over complicated subjects and employs facile generalizations in almost every sentence. Readers rarely have to think, and are spared all of the difficulty and tedium typically involved in reading old-fashioned books." As long as we stick to simple operations like setting the clocks on our VCRs, this is not so bad. But as we try to do more ambitious things with our technologies, we inevitably run into the problem of:

METAPHOR SHEAR

I began using Microsoft Word as soon as the first version was released around 1985. After some initial hassles I found it to be a better tool than its competition. I wrote a lot of stuff in early versions of Word, storing it all on floppies, and transferred the contents of all my floppies to my first hard drive, which I acquired around 1987. As new versions of Word came out, I faithfully upgraded, reasoning that as a writer it made sense for me to spend a certain amount of money on tools.

Sometime in the mid-1990s I attempted to open one of my old, circa-1985 Word documents using the version of Word then current: 6.0. It didn't work. Word 6.0 did not recognize a document created by an earlier version of itself. By opening it as a text file, I was able to recover the sequences of letters that made up the text of the document. My words were still there. But the formatting had been run through a log chipper—the words I'd written were interrupted by spates of empty rectangular boxes and gibberish.

Now, in the context of a business (the chief market for

Word) this sort of thing is only an annoyance—one of the routine hassles that go along with using computers. It's easy to buy little file converter programs that will take care of this problem. But if you are a writer whose career is words, whose professional identity is a corpus of written documents, this kind of thing is extremely disquieting. There are very few fixed assumptions in my line of work, but one of them is that once you have written a word, it is written, and cannot be unwritten. The ink stains the paper, the chisel cuts the stone, the stylus marks the clay, and something has irrevocably happened. (My brother-in-law is a theologian who reads 3250-year-old cuneiform tablets—he can recognize the handwriting of particular scribes and identify them by name.) But word-processing software—particularly the sort that employs special, complex file formats—has the eldritch power to unwrite things. A small change in file formats, or a few twiddled bits, and months' or years' literary output can cease to exist.

Now this was technically a fault in the application (Word 6.0 for the Macintosh) not the operating system (MacOS 7.-something) and so the initial target of my annoyance was the people who were responsible for Word. But. On the other hand, I could have chosen the "Save as Text" option in Word and saved all of my documents as simple "telegrams," and this problem would not have arisen. Instead I had allowed myself to be seduced by all of those flashy formatting options that hadn't even existed until GUIs had come along to make them practicable. I had gotten into the habit of using them to make my documents look pretty (perhaps prettier than they deserved to look; all of the old documents on those floppies turned out to be more or less crap). Now I was paying the price for that self-indulgence. Technology had moved on and found ways to make my documents look even prettier, and the consequence of it was that all old ugly documents had ceased to exist.

It was—if you'll pardon me for a moment's strange little fantasy—as if I'd gone to stay at some resort, some exquisitely designed and art-directed hotel, placing myself in the hands of past masters of the Sensorial Interface, and had sat down in my room and written a story in ballpoint pen on a yellow legal pad, and when I returned from dinner, discovered that the maid had taken my work away and left behind in its place a quill pen and a stack of fine parchment—explaining that the room looked ever so much finer this way, and it was all part of a routine upgrade. But written on these sheets of paper, in flawless penmanship, were long sequences of words chosen at random from the dictionary. Appalling, sure, but I couldn't really lodge a complaint with the management, because by staying at this resort I had given my consent to it. I had surrendered my Morlock credentials and become an Eloi.

LINUX

During the late 1980s and early 1990s I spent a lot of time programming Macintoshes, and eventually decided to fork over several hundred dollars for an Apple product called the Macintosh Programmer's Workshop, or MPW. MPW had competitors, but it was unquestionably the premier software development system for the Mac. It was what Apple's own engineers used to write Macintosh code. Given that MacOS was far more technologically advanced, at the time, than its competition, and that Linux did not even exist yet, and given that this was the actual program used by Apple's world-class team of creative engineers, I had high expectations. It arrived on a stack of floppy disks about a foot high, and so there was plenty of time for my excitement to build during the endless installation process. The first time I launched MPW, I was probably expecting some kind of touchy-feely multimedia showcase. Instead it was austere, almost to the point of being intimidating. It was a scrolling window into which you could type simple, unformatted text. The system would then interpret these lines of text as commands and try to execute them.

In the Beginning . . .

It was, in other words, a glass teletype running a command line interface. It came with all sorts of cryptic but powerful commands, which could be invoked by typing their names, and which I learned to use only gradually. It was not until a few years later, when I began messing around with Unix, that I understood that the command line interface embodied in MPW was a re-creation of Unix.

The first thing that Apple's hackers had done when they'd gotten the MacOS up and running—probably even before they'd gotten it up and running—was to re-create the Unix interface, so that they would be able to get some useful work done. At the time, I simply couldn't get my mind around this, but, apparently as far as Apple's hackers were concerned, the Mac's vaunted graphical user interface was an impediment, something to be circumvented before the little toaster even came out onto the market.

Even before my PowerBook crashed and obliterated my big file in July 1995, there had been danger signs. An old college buddy of mine, who starts and runs high-tech companies in Boston, had developed a commercial product using Macintoshes as the front end. Basically the Macs were high-performance graphics terminals, chosen for their sweet user interface, giving users access to a large database of graphical information stored on a network of much more powerful, but less user-friendly, computers. This fellow was the second person who turned me on to Macintoshes, by the way, and through the mid-1980s we had shared the thrill of being high-tech cognoscenti, using superior Apple technology in a world of DOS-using knuckleheads. Early versions of my friend's system had worked well, he told me, but when several machines joined the network, mysterious crashes began to occur; sometimes the whole network would just freeze. It was one of those bugs that could not be reproduced easily. Finally they figured out that these network crashes were triggered whenever a user, scanning the menus for a particular item, held down the mouse button for more than a couple of seconds.

Fundamentally, the MacOS could only do one thing at a time. Drawing a menu on the screen is one thing. So when a menu was pulled down, the Macintosh was not capable of doing anything else until that indecisive user released the button.

This is not such a bad thing in a single-user, single-process machine (although it's a fairly bad thing), but it's no good in a machine that is on a network, because being on a network implies some kind of continual low-level interaction with other machines. By failing to respond to the network, the Mac caused a network-wide crash.

In order to work with other computers, and with networks, and with various different types of hardware, an OS must be incomparably more complicated and powerful than either MS-DOS or the original MacOS. The only way of connecting to the Internet that's worth taking seriously is PPP, the Point-to-Point Protocol, which (never mind the details) makes your computer—temporarily—a full-fledged member of the global Internet, with its own unique address, and various privileges, powers, and responsibilities appertaining thereunto. Technically it means your machine is running the TCP/IP protocol, which, to make a long story short, revolves around sending packets of data back and forth, in no particular order, and at unpredictable times, according to a clever and elegant set of rules.

But sending a packet of data is one thing, and so an OS that can only do one thing at a time cannot simultaneously be part of the Internet and do anything else. When TCP/IP was invented, running it was an honor reserved for Serious Computers—mainframes and high-powered minicomputers used in technical and commercial settings—and so the protocol is engineered around the assumption that every computer using it is a serious machine, capable of doing many things at once. Not to put too fine a point on it, a Unix machine. Neither MacOS nor MS-DOS was originally built with that in mind, and so when the Internet got hot, radical changes had to be made.

When my PowerBook broke my heart, and when Word stopped recognizing my old files, I jumped to Unix. The obvious alternative to MacOS would have been Windows. I didn't really have anything against Microsoft or Windows. But it was pretty obvious, now, that old PC operating systems were overreaching and showing the strain and, perhaps, were best avoided until they had learned to walk and chew gum at the same time.

The changeover took place on a particular day in the summer of 1995. I had been in San Francisco for a couple of weeks, using my PowerBook to work on a document. The document was too big to fit onto a single floppy, and so I hadn't made a backup since leaving home. The PowerBook crashed and wiped out the entire file.

It happened just as I was on my way out the door to visit a company called Electric Communities, which in those days was in Los Altos. I took my PowerBook with me. My friends at Electric Communities were Mac users who had all sorts of utility software for unerasing files and recovering from disk crashes, and I was certain I could get most of the file back.

As it turned out, two different Mac crash recovery utilities were unable to find any trace that my file had ever existed. It was completely and systematically wiped out. We went through that hard disk block by block and found disjointed fragments of countless old, discarded, forgotten files, but none of what I wanted. The metaphor shear was especially brutal that day. It was sort of like watching the girl you've been in love with for ten years get killed in a car wreck, then attending her autopsy, and learning that underneath the clothes and makeup she was just flesh and blood.

I must have been reeling around the offices of Electric Communities in some kind of primal Jungian fugue, because at this moment three weirdly synchronistic things happened.

(1) Randy Farmer, a cofounder of the company, came in for a quick visit along with his family—he was

recovering from back surgery at the time. He had some hot gossip: "Windows 95 mastered today." What this meant was that Microsoft's new operating system had, on this day, been placed on a special compact disk known as a golden master, which would be used to stamp out a jintillion copies in preparation for its thunderous release a few weeks later. This news was received peevishly by the staff of Electric Communities, including one whose office door was plastered with the usual assortment of cartoons and novelties, e.g.:

(2) A copy of a *Dilbert* cartoon in which Dilbert, the long-suffering corporate software engineer, encounters a portly, bearded, hairy man of a certain age— a bit like Santa Claus, but darker, with a certain edge about him. Dilbert recognizes this man, based upon his appearance and affect, as a Unix hacker, and reacts with a certain mixture of nervousness, awe, and hostility. Dilbert jabs weakly at the disturbing interloper for a couple of frames; the Unix hacker listens with a kind of infuriating, beatific calm, then, in the last frame, reaches into his pocket. "Here's a nickel, kid," he says, "go buy yourself a real computer."

(3) The owner of the door, and the cartoon, was one Doug Barnes. Barnes was known to harbor certain heretical opinions on the subject of operating systems. Unlike most Bay Area techies who revered

the Macintosh, considering it to be a true hacker's machine, Barnes was fond of pointing out that the Mac, with its hermetically sealed architecture, was actually hostile to hackers, who are prone to tinkering and dogmatic about openness. By contrast, the IBM-compatible line of machines, which can easily be taken apart and plugged back together, was much more hackable.

So when I got home I began messing around with Linux, which is one of many, many different concrete implementations of the abstract, Platonic ideal called Unix. I was not looking forward to changing over to a new OS, because my credit cards were still smoking from all the money I'd spent on Mac hardware over the years. But Linux's great virtue was, and is, that it would run on exactly the same sort of hardware as the Microsoft OSes—which is to say, the cheapest hardware in existence. As if to demonstrate why this was a great idea, I was, within a week or two of returning home, able to get my hands on a then-decent computer (a 33 MIL 486 box) for free, because I knew a guy who worked in an office where they were simply being thrown away. Once I got it home, I yanked the hood off, stuck my hands in, and began switching cards around. If something didn't work, I went to a used-computer outlet and pawed through a bin full of components and bought a new card for a few bucks.

The availability of all this cheap but effective hardware was an unintended consequence of decisions that had been made more than a decade earlier by IBM and Microsoft. When Windows came out, and brought the GUI to a much larger market, the hardware regime changed: the cost of color video cards and high-resolution monitors began to drop, and is dropping still. This free-for-all approach to hardware meant that Windows was unavoidably clunky compared to MacOS. But the GUI brought computing to such a vast audience that volume went way up and prices collapsed. Meanwhile Apple, which so badly wanted a clean, integrated OS with video neatly integrated into processing hardware, had fallen far behind in market share, at least partly because their beautiful hardware cost so much.

But the price that we Mac owners had to pay for superior aesthetics and engineering was not merely a financial one. There was a cultural price too, stemming from the fact that we couldn't open up the hood and mess around with it. Doug Barnes was right. Apple, in spite of its reputation as the machine of choice of scruffy, creative hacker types, had actually created a machine that discouraged hacking, while Microsoft, viewed as a technological laggard and copycat, had created a vast, disorderly parts bazaar—a primordial soup that eventually self-assembled into Linux.

THE HOLE HAWG OF
OPERATING SYSTEMS

Unix has always lurked provocatively in the background of the operating system wars, like the Russian Army. Most people know it only by reputation, and its reputation, as the *Dilbert* cartoon suggests, is mixed. But everyone seems to agree that if it could only get its act together and stop surrendering vast tracts of rich agricultural land and hundreds of thousands of prisoners of war to the onrushing invaders, it could stomp them (and all other opposition) flat.

It is difficult to explain how Unix has earned this respect without going into mind-smashing technical detail. Perhaps the gist of it can be explained by telling a story about drills.

The Hole Hawg is a drill made by the Milwaukee Tool Company. If you look in a typical hardware store you may find smaller Milwaukee drills, but not the Hole Hawg, which is too powerful and too expensive for homeowners. The Hole Hawg does not have the pistol-like design of a cheap homeowner's drill. It is a cube of solid metal with a handle sticking out of one face and a chuck mounted in another. The cube contains a disconcertingly potent electric motor. You can hold the handle and operate the trigger with your index finger, but unless you are exceptionally strong, you cannot control the weight of the Hole Hawg with one hand; it is a two-hander all the way. In order to fight off the counter-torque of the Hole Hawg, you use a separate handle (provided), which you screw into one side of the iron cube or the other depending on whether you are using your left or right hand to operate the trigger. This handle is not a sleek, ergonomically designed item as it would be in a homeowner's drill. It is simply a foot-long chunk of regular galvanized pipe, threaded on one end, with a black rubber handle on the other. If you lose it, you just go to the local plumbing supply store and buy another chunk of pipe.

During the eighties I did some construction work. One day, another worker leaned a ladder against the outside of the building that we were putting up, climbed up to the second-story level, and used the Hole Hawg to drill a hole through the exterior wall. At some point, the drill bit caught in the wall. The Hole Hawg, following its one and only imperative, kept going. It spun the worker's body around like a rag doll, causing him to knock his own ladder down. Fortunately he kept his grip on the Hole Hawg, which remained lodged in the wall, and he simply dangled from it and shouted for help until someone came along and reinstated the ladder.

I myself used a Hole Hawg to drill many holes through studs, which it did as a blender chops cabbage. I also used it to cut a few six-inch-diameter holes through an old lath-and-plaster ceiling. I chucked in a new hole saw, went up to the second story, reached down between the newly installed floor joists, and began to cut through the first-floor ceiling below. Where my homeowner's drill had labored and whined to spin the huge bit around, and had stalled at the slightest obstruction, the Hole Hawg rotated with the stupid consistency of a spinning planet. When the hole saw seized up, the Hole Hawg spun itself and me around, and crushed one of my hands between the steel pipe handle and a joist, producing a few lacerations, each surrounded by a wide corona of deeply bruised flesh. It also bent the hole saw itself, though not so badly that I couldn't use it. After a few such run-ins, when I got ready to use the Hole Hawg, my heart actually began to pound with atavistic terror.

But I never blamed the Hole Hawg; I blamed myself. The Hole Hawg is dangerous because it does exactly what you tell it to. It is not bound by the physical limitations that are inherent in a cheap drill, and neither is it limited by safety interlocks that might be built into a homeowner's product by a liability-conscious manufacturer. The danger lies not in the machine itself but in the user's failure to envision the full consequences of the instructions he gives to it.

A smaller tool is dangerous too, but for a completely different reason: it tries to do what you tell it to, and fails in some way that is unpredictable and almost always undesirable. But the Hole Hawg is like the genie of the ancient fairy tales, who carries out his master's instructions literally and precisely and with unlimited power, often with disastrous, unforeseen consequences.

Pre–Hole Hawg, I used to examine the drill selection in hardware stores with what I thought was a judicious eye, scorning the smaller low-end models and hefting the big expensive ones appreciatively, wishing I could afford one of them babies. Now I view them all with such contempt that I do not even consider them to be real drills—merely scaled-up toys designed to exploit the self-delusional tendencies of soft-handed homeowners who want to believe that they have purchased an actual tool. Their plastic casings, carefully designed and focus-group tested to convey a feeling of solidity and power, seem disgustingly flimsy and cheap to me, and I am ashamed that I was ever bamboozled into buying such knicknacks.

It is not hard to imagine what the world would look like to someone who had been raised by contractors and who had never used any drill other than a Hole Hawg. Such a person, presented with the best and most expensive hardware-store drill, would not even recognize it as such. He might instead misidentify it as a child's toy, or some kind of motorized screwdriver. If a salesperson or a deluded homeowner referred to it as a drill, he would laugh and tell them that they were mistaken—they simply had their terminology wrong. His interlocutor would go away irritated, probably feeling rather defensive about his basement full of cheap, dangerous, flashy, colorful tools.

Unix is the Hole Hawg of operating systems,* and Unix hackers—like Doug Barnes and the guy in the *Dilbert* cartoon and many of the other people who populate Silicon Valley—are like contractors' sons who grew up using only Hole Hawgs. They might use Apple/Microsoft OSes to write letters, play video games, or balance their checkbooks, but they cannot really bring themselves to take these operating systems seriously.

---

* Dr. Myhrvold of Microsoft has laid down his dinosaur pick, risen to the challenge, and countered with a trenchant drill analogy of his own that spins in the opposite direction, as it were. His drill analogy is probably, in the end, better than mine. I will not present it here, because a public drill analogy duel would present a ridiculous and undignified spectacle. Here are some excerpts:

"There is a silly romanticism that a more primitive instrument that requires lots of skill from the operator must somehow be more powerful. It's usually bullshit. . . ."

"An important reason that Linux has become interesting is that the Internet has caused a temporarily retro phase when interesting programs are suddenly very unsophisticated. Apache, or an NNTP server, is very simple software that does not require much of an OS. The same is true for many other web-oriented tasks. Linux is fine for these."

THE ORAL TRADITION

Unix is hard to learn. The process of learning it is one of multiple small epiphanies. Typically you are just on the verge of inventing some necessary tool or utility when you realize that someone else has already invented it, and built it in, and this explains some odd file or directory or command that you have noticed but never really understood before.

For example, there is a command (a small program, part of the OS) called "whoami," which enables you to ask the computer who it thinks you are. On a Unix machine, you are always logged in under some name—possibly even your own! What files you may work with, and what software you may use, depends on your identity. When I started out using Linux, I was on a nonnetworked machine in my basement, with only one user account, and so when I became aware of the whoami command it struck me as ludicrous. But once you are logged in as one person, you can temporarily switch over to a pseudonym in order to access different files. If your machine is on the Internet, you can log on to other computers, provided you have a user name and a password. At that point the distant machine becomes no different in practice from the one right in front of you. These changes in identity and location can easily become nested inside each other, many layers deep, even if you aren't doing anything nefarious. Once you have forgotten who and where you are, the whoami command is indispensible. I use it all the time.

The file systems of Unix machines all have the same general structure. On your flimsy operating systems, you can create directories (folders) and give them names like Frodo or My Stuff and put them pretty much anywhere you like. But under Unix the highest level—the root— of the filesystem is always designated with the single character "/" and it always contains the same set of top-level directories:

/usr /etc /var /bin /proc /boot /home /root /sbin /dev /lib /tmp

Each of these directories typically has its own distinct structure of subdirectories. Note the obsessive use of abbreviations and avoidance of capital letters; this is a system invented by people to whom repetitive stress disorder is what black lung is to miners. Long names get worn down to three- or four-letter nubbins, like stones smoothed by a river.

This is not the place to try to explain why each of the above directories exists and what is contained in them. At first it all seems obscure; worse, it seems deliberately obscure. When I started using Linux, I was accustomed to being able to create directories wherever I wanted and to give them whatever names struck my fancy. Under Unix you are free to do that, of course (you are free to do anything), but as you gain experience with the system you come to understand that the directories listed above were created for the best of reasons and that your life will be much easier if you follow along. (Within /home, by the way, you have pretty much unlimited freedom.)

After this kind of thing has happened several hundred or thousand times, the hacker understands why Unix is the way it is, and agrees that it wouldn't be the same any other way. It is this sort of acculturation that gives Unix hackers their confidence in the system and the attitude of calm, unshakable, annoying superiority captured in the *Dilbert* cartoon. Windows 95 and MacOS are products, contrived by engineers in the service of specific companies. Unix, by contrast, is not so much a product as it is a painstakingly compiled oral history of the hacker subculture. It is our *Gilgamesh* epic.

What made old epics like *Gilgamesh* so powerful and so long-lived was that they were living bodies of narrative that many people knew by heart, and told over and over again—making their own personal embellishments whenever it struck their fancy. The bad embellishments were shouted down, the good ones picked up by others, polished, improved and, over time, incorporated into the story. Likewise, Unix is known, loved, and understood by so many hackers that it can be re-created from scratch whenever someone needs it. This is very difficult to understand for people who are accustomed to thinking of OSes as things that absolutely have to be created by a company and bought.

Many hackers have launched more or less successful reimplementations of the Unix ideal. Each one brings in new embellishments. Some of them die out quickly, some are merged with similar, parallel innovations created by different hackers attacking the same problem, others still are embraced and adopted into the epic. Thus Unix has slowly accreted around a simple kernel and acquired a kind of complexity and asymmetry that is organic, like the roots of a tree, or the branchings of a coronary artery. Understanding it is more like anatomy than physics.

For at least a year, prior to my adoption of Linux, I had been hearing about it. Credible, well-informed people kept telling me that a bunch of hackers had got together an implentation of Unix that could be downloaded, free of charge, from the Internet. For a long time I could not bring myself to take the notion seriously. It was like hearing rumors that a group of model-rocket enthusiasts had created a completely functional Saturn V by exchanging blueprints on the Net and mailing valves and flanges to each other.

But it's true. Credit for Linux generally goes to its human namesake, one Linus Torvalds, a Finn who got the whole thing rolling in 1991 when he used some of the GNU tools to write the beginnings of a Unix kernel that could run on PC-compatible hardware. And indeed Torvalds deserves all the credit he has ever gotten, and a whole lot more. But he could not have made it happen by himself, any more than Richard Stallman could have. To write code at all, Torvalds had to have cheap but powerful development tools, and these he got from Stallman's GNU project.

And he had to have cheap hardware on which to write that code. Cheap hardware is a much harder thing to arrange than cheap software. A single person (Stallman) can write software and put it up on the Net for free, but in order to make hardware it's necessary to have a whole industrial infrastructure, which is not cheap by any stretch of the imagination. Really the only way to make hardware cheap is to punch out an incredible number of copies of it, so that the unit cost eventually drops. For reasons already explained, Apple had no desire to see the cost of hardware drop. The only reason Torvalds had cheap hardware was Microsoft.

Microsoft refused to go into the hardware business, insisted on making its software run on hardware that anyone could build, and thereby created the market conditions that allowed hardware prices to plummet. In trying to understand the Linux phenomenon, then, we have to look not to a single innovator but to a sort of bizarre Trinity: Linus Torvalds, Richard Stallman, and Bill Gates. Take away any of these three and Linux would not exist.

OS SHOCK

Young Americans who leave their great big homogeneous country and visit some other part of the world typically go through several stages of culture shock: first, dumb wide-eyed astonishment. Then, a tentative engagement with the new country's manners, cuisine, public transit systems, and toilets, leading to a brief period of fatuous confidence that they are instant experts on the new country. As the visit wears on, homesickness begins to set in, and the traveler begins to appreciate, for the first time, how much he or she took for granted at home. At the same time it begins to seem obvious that many of one's own cultures and traditions are essentially arbitrary and could have been different; driving on the right side of the road, for example. When the traveler returns home and takes stock of the experience, he or she may have learned a good deal more about America than about the country they went to visit.

For the same reasons, Linux is worth trying. It is a strange country indeed, but you don't have to live there; a brief sojourn suffices to give some flavor of the place and—more importantly—to lay bare everything that is taken for granted, and all that could have been done differently, under Windows or MacOS.

You can't try it unless you install it. With any other OS, installing it would be a straightforward transaction: in exchange for money, some company would give you a CD-ROM, and you would be on your way. But a lot is subsumed in that kind of transaction, and it has to be gone through and picked apart.

We like plain dealings and straightforward transactions in America. If you go to Egypt and, say, take a taxi somewhere, you become a part of the taxi driver's life; he refuses to take your money because it would demean your friendship, he follows you around town, and weeps hot tears when you get in some other guy's taxi. You end up meeting his kids at some point and have to devote all sorts of ingenuity to finding some way to compensate him without insulting his honor. It is exhausting. Sometimes you just want a simple Manhattan-style taxi ride.

But in order to have an American-style setup, where you can just go out and hail a taxi and be on your way, there must exist a whole hidden apparatus of medallions, inspectors, commissions, and so forth—which is fine as long as taxis are cheap and you can always get one. When the system fails to work in some way, it is mysterious and infuriating and turns otherwise reasonable people into conspiracy theorists. But when the Egyptian system breaks down, it breaks down transparently. You can't get a taxi, but your driver's nephew will show up, on foot, to explain the problem and apologize.

Microsoft and Apple do things the Manhattan way, with vast complexity hidden behind a wall of interface. Linux does things the Egypt way, with vast complexity strewn about all over the landscape. If you've just flown in from Manhattan, your first impulse will be to throw up your hands and say, "For crying out loud! Will you people get a grip on yourselves!?" But this does not make friends in Linux-land any better than it would in Egypt.

You can suck Linux right out of the air, as it were, by downloading the right files and putting them in the right places, but there probably are not more than a few hundred people in the world who could create a functioning Linux system in that way. What you really need is a distribution of Linux, which means a prepackaged set of files. But distributions are a separate thing from Linux per se.

Linux per se is not a specific set of ones and zeroes, but a self-organizing Net subculture. The end result of its collective lucubrations is a vast body of source code, almost all written in C (the dominant computer programming language). "Source code" just means a computer program as typed in and edited by some hacker. If it's in C, the file name will probably have .c or .cpp on the end of it, depending on which dialect was used; if it's in some other language, it will have some other suffix. Frequently these sorts of files can be found in a directory with the name /src, which is the hacker's Hebraic abbreviation of "source."

Source files are useless to your computer, and of little

interest to most users, but they are of gigantic cultural and political significance, because Microsoft and Apple keep them secret while Linux makes them public. They are the family jewels. They are the sort of thing that in Hollywood thrillers is used as a McGuffin: the plutonium bomb core, the top-secret blueprints, the suitcase of bearer bonds, the reel of microfilm. If the source files for Windows or MacOS were made public on the Net, then those OSes would become free, like Linux—only not as good, because no one would be around to fix bugs and answer questions. Linux is "open source" software, meaning simply, anyone can get copies of its source code files.

Your computer doesn't want source code any more than you do; it wants object code. Object code files typically have the suffix .o and are unreadable to all but a few, highly strange humans, because they consist of ones and zeroes. Accordingly, this sort of file commonly shows up in a directory with the name /bin, for "binary."

Source files are simply ASCII text files. ASCII denotes a particular way of encoding letters into bit patterns. In an ASCII file, each character has eight bits all to itself. This creates a potential "alphabet" of 256 distinct characters, in that eight binary digits can form that many unique patterns. In practice, of course, we tend to limit ourselves to the familiar letters and digits. The bit-patterns used to represent those letters and digits are the same ones that were physically punched into the paper tape by my high school teletype, which in turn were the same ones used by the telegraph industry for decades previously. ASCII text files, in other words, are telegrams, and as such they have no typographical frills. But for the same reason they are eternal, because the code never changes, and universal, because every text-editing and word-processing software ever written knows about this code.

Therefore just about any software can be used to create, edit, and read source code files. Object code files, then, are created from these source files by a piece of software called a compiler, and forged into a working application by another piece of software called a linker.

The triad of editor, compiler, and linker, taken together, form the core of a software development system. Now, it is possible to spend a lot of money on shrink-wrapped development systems with lovely graphical user interfaces and various ergonomic enhancements. In some cases it might even be a good and reasonable way to spend money. But on this side of the road, as it were, the very best software is usually the free stuff. Editor, compiler, and linker are to hackers what ponies, stirrups, and archery sets were to the Mongols. Hackers live in the saddle and hack on their own tools, even while they are using them, to create new applications. It is quite inconceivable that superior hacking tools could have been created from a blank sheet of paper by product engineers. Even if they are the brightest engineers in the world, they are simply outnumbered.

In the GNU/Linux world, there are two major text-editing programs: the minimalist vi (known in some implementations as elvis) and the maximalist emacs. I use emacs, which might be thought of as a thermonuclear word processor. It was created by Richard Stallman;

enough said. It is written in Lisp, which is the only computer language that is beautiful. It is colossal, and yet it
only edits straight ASCII text files, which is to say, no
fonts, no boldface, no underlining. In other words, the
engineer-hours that, in the case of Microsoft Word, were
devoted to features like mail merge, and the ability to
embed feature-length motion pictures in corporate memo-
randa, were, in the case of emacs, focused with maniacal
intensity on the deceptively simple-seeming problem of editing text. If you are a professional writer—i.e., if someone
else is getting paid to worry about how your words are
formatted and printed—emacs outshines all other editing
software in approximately the same way that the noonday
sun does the stars. It is not just bigger and brighter; it
simply makes everything else vanish. For page layout and
printing you can use TeX: a vast corpus of typesetting lore
written in C and also available on the Net for free.

I could say a lot about emacs and TeX, but right now
I am trying to tell a story about how to actually install
Linux on your machine. The hard-core survivalist approach would be to download an editor like emacs, and
the GNU Tools—the compiler and linker—which are
polished and excellent to the same degree as emacs.
Equipped with these, one would be able to start downloading ASCII source code files (/src) and compiling
them into binary object code files (/bin) that would run
on the machine. But in order to even arrive at this
point—to get emacs running, for example—you have to
have Linux actually up and running on your machine.
And even a minimal Linux operating system requires

thousands of binary files all acting in concert, and arranged and linked together just so.

Several entities have therefore taken it upon themselves to create "distributions" of Linux. If I may extend the Egypt analogy slightly, these entities are a bit like tour guides who meet you at the airport, who speak your language, and who help guide you through the initial culture shock. If you are an Egyptian, of course, you see it the other way; tour guides exist to keep brutish outlanders from traipsing through your mosques and asking you the same questions over and over and over again.*

Some of these tour guides are commercial organizations, such as Red Hat Software, which makes a Linux distribution called Red Hat that has a relatively commercial sheen to it. In most cases you put a Red Hat CD-ROM into your PC and reboot and it handles the rest. Just as a tour guide in Egypt will expect some sort of compensation for his services, commercial distributions need to be paid for. In most cases, they cost almost nothing and are well worth it.

I use a distribution called Debian† (the word is a contrac-

---

* In any exotic country, the best tour guide is a native who is fluent in English. Eric S. Raymond is an eminent open-source hacker who has become the foremost anthropologist of the open-source tribe. He has an ongoing series of papers, available on the web. The first and best-known is "The Cathedral and the Bazaar." The second is "Homesteading the Noosphere." Others are planned. Probably the most reliable way of finding these papers is to visit Raymond's website at www.tuxedo.org/~esr/

† Again, the full Stallman-compliant term for this would be "Debian GNU/ Linux." This nomenclature is an implicit way of reminding us of something that I have here tried to state explicitly, namely that none of this would exist without GNU.

tion of "Deborah" and "Ian") which is noncommercial. It is organized (or perhaps I should say "it has organized itself") along the same lines as Linux in general, which is to say that it consists of volunteers who collaborate over the Net, each responsible for looking after a different chunk of the system. These people have broken Linux down into a number of packages, which are compressed files that can be downloaded to an already functioning Debian Linux system, then opened up and unpacked using a free installer application. Of course, as such, Debian has no commercial arm—no distribution mechanism. You can download all Debian packages over the Net, but most people will want to have them on a CD-ROM. Several different companies have taken it upon themselves to decoct all of the current Debian packages onto CD-ROMs and then sell them. I buy mine from Linux Systems Labs. The cost for a three-disk set, containing Debian in its entirety, is less than three dollars. But (and this is an important distinction) not a single penny of that three dollars is going to any of the coders who created Linux, nor to the Debian packagers. It goes to Linux Systems Labs and it pays not for the software or the packages but for the cost of stamping out the CD-ROMs.

Every Linux distribution embodies some more or less clever hack for circumventing the normal boot process and causing your computer, when it is turned on, to organize itself not as a PC running Windows but as a "host" running Unix. This is slightly alarming the first time you see it, but completely harmless. When a PC boots up, it goes through a little self-test routine, taking an inventory of available disks and memory, and then begins looking around for a disk to boot up from. In any normal Windows computer that disk will be a hard drive. But if you have your system configured right, it will look first for a floppy or CD-ROM disk, and boot from that if one is available.

Linux exploits this chink in the defenses. Your computer notices a bootable disk in the floppy or CD-ROM drive, loads in some object code from that disk, and blindly begins to execute it. But this is not Microsoft or Apple code, this is Linux code, and so at this point your computer begins to behave very differently from what you are accustomed to. Cryptic messages began to scroll up the screen. If you had booted a commercial OS, you would, at this point, be seeing a "Welcome to MacOS" cartoon, or a screen filled with clouds in a blue sky and a Windows logo. But under Linux you get a long telegram printed in stark white letters on a black screen. There is no "Welcome!" message. Most of the telegram has the semi-inscrutable menace of graffiti tags.

Dec 14 15:04:15 theRev syslogd 1.3-3#17: restart. Dec 14
15:04:15 theRev kernel: klogd 1.3-3, log source =
/proc/kmsg started. Dec 14 15:04:15 theRev kernel: Loaded
3535 symbols from /System.map. Dec 14 15:04:15
theRev kernel: Symbols match kernel version 2.0.30. Dec
14 15:04:15 theRev kernel: No module symbols loaded.
Dec 14 15:04:15 theRev kernel: Intel MultiProcessor
Specification v1.4 Dec 14 15:04:15 theRev kernel:
Virtual Wire compatibility mode. Dec 14 15:04:15 theRev

In the Beginning . . .

kernel: OEM ID: INTEL Product ID: 440FX APIC at: 0xFEE00000 Dec 14 15:04:15 theRev kernel: Processor #0 Pentium(tm) Pro APIC version 17 Dec 14 15:04:15 theRev kernel: Processor #1 Pentium(tm) Pro APIC version 17 Dec 14 15:04:15 theRev kernel: I/O APIC #2 Version 17 at 0xFEC00000. Dec 14 15:04:15 theRev kernel: Processors: 2 Dec 14 15:04:15 theRev kernel: Console: 16 point font, 400 scans Dec 14 15:04:15 theRev kernel: Console: colour VGA+ 80x25, 1 virtual console (max 63) Dec 14 15:04:15 theRev kernel: pcibios_init : BIOS32 Service Directory structure at 0x000fdb70 Dec 14 15:04:15 theRev kernel: pcibios_init : BIOS32 Service Directory entry at 0xfdb80 Dec 14 15:04:15 theRev kernel: pcibios_init : PCI BIOS revision 2.10 entry at 0xfdba1 Dec 14 15:04:15 theRev kernel: Probing PCI hardware. Dec 14 15:04:15 theRev kernel: Warning : Unknown PCI device (10b7:9001). Please read include/linux/pci.h Dec 14 15:04:15 theRev kernel: Calibrating delay loop... Ok - 179.40 BogoMIPS Dec 14 15:04:15 theRev kernel: Memory: 64268k/66556k available (700k kernel code, 384k reserved, 1204k data) Dec 14 15:04:15 theRev kernel: Swansea University Computer Society NET3.035 for Linux 2.0 Dec 14 15:04:15 theRev kernel: NET3: Unix domain sockets 0.13 for Linux NET3.035. Dec 14 15:04:15 theRev kernel: Swansea University Computer Society TCP/IP for NET3.034 Dec 14 15:04:15 theRev kernel: IP Protocols: ICMP, UDP, TCP Dec 14 15:04:15 theRev kernel: Checking 386/387 coupling... Ok, fpu using exception 16 error reporting. Dec 14 15:04:15 theRev kernel: Checking 'hlt' instruction...

Ok. Dec 14 15:04:15 theRev kernel: Linux version
2.0.30 (root@theRev) (gcc version 2.7.2.1) #15 Fri Mar 27
16:37:24 PST 1998 Dec 14 15:04:15 theRev kernel:
Booting processor 1 stack 00002000: Calibrating delay
loop.. ok -179.40 BogoMIPS Dec 14 15:04:15 theRev
kernel: Total of 2 processors activated (358.81
BogoMIPS). Dec 14 15:04:15 theRev kernel: Serial
driver version 4.13 with no serial options enabled Dec
14 15:04:15 theRev kernel: tty00 at 0x03f8 (irq = 4)
is a 16550A Dec 14 15:04:15 theRev kernel: tty01 at
0x02f8 (irq = 3) is a 16550A Dec 14 15:04:15 theRev
kernel: 1p1 at 0x0378, (polling) Dec 14 15:04:15 theRev
kernel: PS/2 auxiliary pointing device detected -- driver
installed. Dec 14 15:04:15 theRev kernel: Real Time
Clock Driver v1.07 Dec 14 15:04:15 theRev kernel:
loop: registered device at major 7 Dec 14 15:04:15
theRev kernel: ide: i82371 PIIX (Triton) on PCI bus 0
function 57 Dec 14 15:04:15 theRev kernel: ide0:
BM-DMA at 0xffa0-0xffa7 Dec 14 15:04:15 theRev
kernel: ide1: BM-DMA at 0xffa8-0xffaf Dec 14 15:04:15
theRev kernel: hda: Conner Peripherals 1275MB -
CFS1275A, 1219MB w/64kB Cache, LBA, CHS=619/64/
63 Dec 14 15:04:15 theRev kernel: hdb: Maxtor
84320A5, 4119MB w/256kB Cache, LBA, CHS=8928/
15/63, DMA Dec 14 15:04:15 theRev kernel: hdc: ,
ATAPI CDROM drive Dec 15 11:58:06 theRev kernel:
ide0 at 0x1f0-0x1f7,0x3f6 on irq 14 Dec 15 11:58:06
theRev kernel: ide1 at 0x170-0x177,0x376 on irq 15 Dec
15 11:58:06 theRev kernel: Floppy drive(s): fd0 is
1.44M Dec 15 11:58:06 theRev kernel: Started kswapd

In the Beginning . . .

v 1.4.2.2 Dec 15 11:58:06 theRev kernel: FDC 0 is a National Semiconductor PC87306 Dec 15 11:58:06 theRev kernel: md driver 0.35 MAX_MD_DEV=4, MAX_REAL=8 Dec 15 11:58:06 theRev kernel: PPP: version 2.2.0 (dynamic channel allocation) Dec 15 11:58:06 theRev kernel: TCP compression code copyright 1989 Regents of the University of California Dec 15 11:58:06 theRev kernel: PPP Dynamic channel allocation code copyright 1995 Caldera, Inc. Dec 15 11:58:06 theRev kernel: PPP line discipline registered. Dec 15 11:58:06 theRev kernel: SLIP: version 0.8.4-NET3.019-NEWTTY (dynamic channels, max=256). Dec 15 11:58:06 theRev kernel: eth0: 3Com 3c900 Boomerang 10Mbps/Combo at 0xef00, 00:60:08:a4:3c:db, IRQ 10 Dec 15 11:58:06 theRev kernel: 8K word-wide RAM 3:5 Rx:Tx split, 10base2 interface. Dec 15 11:58:06 theRev kernel: Enabling bus-master transmits and whole-frame receives. Dec 15 11:58:06 theRev kernel: 3c59x.c:v0.49 1/2/98 Donald Becker http: cesdis.gsfc.nasa.gov/linux/drivers/ vortex.html Dec 15 11:58:06 theRev kernel: Partition check: Dec 15 11:58:06 theRev kernel: hda: hda1 hda2 hda3 Dec 15 11:58:06 theRev kernel: hdb: hdb1 hdb2 Dec 15 11:58:06 theRev kernel: VFS: Mounted root (ext2 filesystem) readonly. Dec 15 11:58:06 theRev kernel: Adding Swap: 16124k swap-space (priority -1) Dec 15 11:58:06 theRev kernel: EXT2-fs warning: maximal mount count reached, running e2fsck is recommended Dec 15 11:58:06 theRev kernel: hdc: media changed Dec 15 11:58:06 theRev kernel: ISO9660 Extensions:

RRIP_1991A Dec 15 11:58:07 theRev syslogd
1.3-3#17: restart. Dec 15 11:58:09 theRev diald[87]: Unable
to open options file /etc/diald/diald.options: No such
file or directory Dec 15 11:58:09 theRev diald[87]: No
device specified. You must have at least one device!
Dec 15 11:58:09 theRev diald[87]: You must define a
connector script (option 'connect'). Dec 15 11:58:09
theRev diald[87]: You must define the remote ip address.
Dec 15 11:58:09 theRev diald[87]: You must define
the local ip address. Dec 15 11:58:09 theRev diald[87]:
Terminating due to damaged reconfigure.

The only parts of this that are readable, for normal people, are the error messages and warnings. And yet it's note-
worthy that Linux doesn't stop, or crash, when it
encounters an error; it spits out a pithy complaint, gives
up on whatever processes were damaged, and keeps on
rolling. This was decidedly not true of the early versions
of Apple and Microsoft OSes, for the simple reason that
an OS that is not capable of walking and chewing gum at
the same time cannot possibly recover from errors. Looking for, and dealing with, errors requires a separate process
running in parallel with the one that has erred. A kind of
superego, if you will, that keeps an eye on all of the others
and jumps in when one goes astray. Now that MacOS and
Windows can do more than one thing at a time, they are
much better at dealing with errors than they used to be,
but they are not even close to Linux or other Unices in
this respect, and their greater complexity has made them
vulnerable to new types of errors.

FALLIBILITY, ATONEMENT, REDEMPTION, TRUST, AND OTHER ARCANE TECHNICAL CONCEPTS

Linux is not capable of having any centrally organized policies dictating how to write error messages and documentation, and so each programmer writes his own. Usually they are in English, even though tons of Linux programmers are Europeans. Frequently they are funny. Always they are honest. If something bad has happened because the software simply isn't finished yet, or because the user screwed something up, this will be stated forthrightly. The command line interface makes it easy for programs to dribble out little comments, warnings, and messages here and there. Even if the application is imploding like a damaged submarine, it can still usually eke out a little S.O.S. message. Sometimes when you finish working with a program and shut it down, you find that it has left behind a series of mild warnings and low-grade error messages in the command line interface window from which you launched it—as if the software were chatting to you about how it was doing the whole time you were working with it.

Documentation, under Linux, comes in the form of man (short for manual) pages. You can access these either through a GUI (xman) or from the command line (man). Here is a sample from the man page for a program called rsh:

Stop signals stop the local rsh process only; this is arguably wrong, but currently hard to fix for reasons too complicated to explain here.

The man pages contain a lot of such material, which reads like the terse mutterings of pilots wrestling with the controls of damaged airplanes. The general feel is of a thousand monumental but obscure struggles seen in the stop-action light of a strobe. Each programmer is dealing with his own obstacles and bugs; he is too busy fixing them, and improving the software, to explain things at great length or to maintain elaborate pretensions.

In practice you hardly ever encounter a serious bug while running Linux. When you do, it is almost always with commercial software (several vendors sell software that runs under Linux, and there is more available each month). The operating system and its fundamental utility programs are too important to contain serious bugs. I have been running Linux every day since late 1995 and have seen many application programs go down in flames, but I have never seen the operating system crash. Never. Not once. There are quite a few Linux systems that have been running continuously and working hard for months or years without needing to be rebooted.

Commercial OSes have to adopt the same official stance toward errors as Communist countries had toward poverty. For doctrinal reasons it was not possible to admit that poverty was a serious problem in Communist countries, because the whole point of Communism was to eradicate poverty. Likewise, commercial OS companies such as Apple and Microsoft can't go around admitting that their software has bugs and that it crashes all the time, any more than Disney can issue press releases stating that Mickey Mouse is an actor in a suit.

This is a problem, because errors do exist and bugs do happen. Every few months Bill Gates tries to demo a new Microsoft product in front of a large audience only to have it blow up in his face. Commercial OS vendors, as a direct consequence of being commercial, are forced to adopt the grossly disingenuous position that bugs are rare aberrations, usually someone else's fault, and therefore not really worth talking about in any detail. This posture, which everyone knows to be absurd, is not limited to press releases and ad campaigns. It informs the whole way these companies do business and relate to their customers. If the documentation were properly written, it would mention bugs, errors, and crashes on every single page. If the on-line help systems that come with these OSes reflected the experiences and concerns of their users, they would largely be devoted to instructions on how to cope with crashes and errors.

But this does not happen. Joint stock corporations are wonderful inventions that have given us many excellent goods and services. They are good at many things. Admitting failure is not one of them. Hell, they can't even admit minor shortcomings.

Of course, this behavior is not as pathological in a corporation as it would be in a human being. Most people, nowadays, understand that corporate press releases are issued for the benefit of the corporation's shareholders and not for the enlightenment of the public. Sometimes the results of this institutional dishonesty can be dreadful, as with tobacco and asbestos. In the case of commercial OS vendors it is nothing of the kind, of course; it is merely annoying.

Some might argue that consumer annoyance, over time, builds up into a kind of hardened plaque that can conceal serious decay, and that honesty might therefore be the best policy in the long run; the jury is still out on this in the operating system market. The business is expanding fast enough that it's still much better to have billions of chronically annoyed customers than millions of happy ones.

Most system administrators I know who work with Windows NT all the time agree that when it hits a snag, it has to be rebooted, and when it gets seriously messed up, the only way to fix it is to reinstall the operating system from scratch. Or at least this is the only way that they know of to fix it, which amounts to the same thing. It is quite possible that the engineers at Microsoft have all sorts of insider knowledge on how to fix the system when it goes awry, but if they do, they do not seem to be getting the message out to any of the actual system administrators I know.

Because Linux is not commercial—because it is, in fact, free, as well as rather difficult to obtain, install, and operate—it does not have to maintain any pretensions as to its reliability. Consequently, it is much more reliable. When something goes wrong with Linux, the error is noticed and loudly discussed right away. Anyone with the requisite technical knowledge can go straight to the source code and point out the source of the error, which is then rapidly fixed by whichever hacker has carved out responsibility for that particular program.

As far as I know, Debian is the only Linux distribution that has its own constitution (http://www.debian.org/devel/constitution), but what really sold me on it was its phenomenal bug database (http://www.debian.org/Bugs), which is a sort of interactive Doomsday Book of error, fallibility, and redemption. It is simplicity itself. When I had a problem with Debian in early January of 1997, I sent in a message describing the problem to submit@bugs.debian.org. My problem was promptly assigned a bug report number (#6518) and a severity level (the available choices being critical, grave, important, normal, fixed, and wishlist) and forwarded to mailing lists where Debian people hang out. Within twenty-four hours I had received five e-mails telling me how to fix the problem:

two from North America, two from Europe, and one from Australia. All of these e-mails gave me the same suggestion, which worked, and made my problem go away. But at the same time, a transcript of this exchange was posted to Debian's bug database, so that if other users had the same problem later, they would be able to search through and find the solution without having to enter a redundant bug report.

Contrast this with the experience that I had when I tried to install Windows NT 4.0 on the very same machine about ten months later, in late 1997. The installation program simply stopped in the middle with no error messages. I went to the Microsoft Support website and tried to perform a search for existing help documents that would address my problem. The search engine was completely nonfunctional; it did nothing at all. It did not even give me a message telling me that it was not working.

Eventually I decided that my motherboard must be at fault; it was of a slightly unusual make and model, and NT did not support as many different motherboards as Linux. I am always looking for excuses, no matter how feeble, to buy new hardware, so I bought a new motherboard that was Windows NT logo-compatible, meaning that the Windows NT logo was printed right on the box. I installed this into my computer and got Linux running right away, then attempted to install Windows NT again. Again, the installation died without any error message or explanation. By this time a couple of weeks had gone by and I thought that perhaps the search engine on the

Microsoft Support website might be up and running. I gave that a try, but it still didn't work.

So I created a new Microsoft support account, then logged on to submit the incident. I supplied my product ID number when asked, and began to follow the instructions on a series of help screens. In other words, I was submitting a bug report just as with the Debian bug tracking system. It's just that the interface was slicker— I was typing my complaint into little text-editing boxes on web forms, doing it all through the GUI, whereas with Debian you send in a simple e-mail "telegram". I knew that when I was finished submitting the bug report, it would become proprietary Microsoft information, and other users wouldn't be able to see it. Many Linux users would refuse to participate in such a scheme on ethical grounds, but I was willing to give it a shot as an experiment. In the end, though I was never able to submit my bug report, because the series of linked web pages that I was filling out eventually led me to a completely blank page: a dead end.

So I went back and clicked on the buttons for "phone support" and eventually was given a Microsoft telephone number. When I dialed this number, I got a series of piercing beeps and a recorded message from the phone company saying, "We're sorry, your call cannot be completed as dialed."

I tried the search page again—it was still completely nonfunctional. Then I tried choosing PPI support (Pay Per Incident) again. This led me through another series of web pages until I dead-ended at one reading: "Notice—there is no Web page matching your request."

I tried it again, and eventually got to a Pay Per Incident screen reading: "OUT OF INCIDENTS. There are no unused incidents left in your account. If you would like to purchase a support incident, click OK—you will then be able to prepay for an incident. . . ." The cost per incident was $95.

The experiment was beginning to seem rather expensive, so I gave up on the PPI approach and decided to have a go at the FAQs posted on Microsoft's website. None of the available FAQs had anything to do with my problem except for one entitled, "I am having some problems installing NT," which appeared to have been written by flacks, not engineers.

So I gave up and still, to this day, have never gotten Windows NT installed on that particular machine. For me, the path of least resistance was simply to use Debian Linux.

In the world of open source software, bug reports are useful information. Making them public is a service to other users, and improves the OS. Making them public systematically is so important that highly intelligent people voluntarily put time and money into running bug databases. In the commercial OS world, however, reporting a bug is a privilege that you have to pay lots of money for. But if you pay for it, it follows that the bug report must be kept confidential—otherwise anyone could get the benefit of your ninety-five bucks!

This is, in other words, another feature of the OS market that simply makes no sense unless you view it in the context of culture. What Microsoft is selling through Pay Per Incident isn't technical support so much as the continued illusion that its customers are engaging in some kind of rational business transaction. It is a sort of routine maintenance fee for the upkeep of the fantasy. If people really wanted a solid OS they would use Linux, and if they really wanted tech support they would find a way to get it; Microsoft's customers must want something else.

As of this writing (January 1999), something like 32,000 bugs have been reported to the Debian Linux bug database. Almost all of them have been fixed a long time ago. There are twelve "critical" bugs still outstanding, of which the oldest was posted seventy-nine days ago. There are twenty outstanding "grave" bugs, of which the oldest is 1166 days old. There are forty-eight "important" bugs, and hundreds of "normal" and less important ones.

Likewise, BeOS (which I'll get to in a minute) has its own bug database (http://www.be.com/developers/bugs) with its own classification system, including such categories as "Not a Bug," "Acknowledged Feature," and "Will Not Fix." Some of the "bugs" here are nothing more than Be hackers blowing off steam, and are classified as "Input Acknowledged." For example, I found one that was posted on December 30, 1998. It's in the middle of a long list of bugs, wedged between one entitled "Mouse working in very strange fashion" and another called "Change of BView frame does not affect, if BView not attached to a BWindow." This one is entitled:

R4: BeOS missing megalomaniacal figurehead to harness and focus developer rage

and it goes like this:

Be Status: Input Acknowledged
BeOS Version: R3.2
Component: unknown

Full Description:

The BeOS needs a megalomaniacal egomaniac sitting on its throne to give it a human character which everyone loves to hate. Without this, the BeOS will languish in the impersonifiable realm of OSs that people can never quite get a handle on. You can judge the success of an OS not by the quality of its features, but by how infamous and disliked the leaders behind them are.

I believe this is a side-effect of developer comraderie [sic] under miserable conditions. After all, misery loves company. I believe that making the BeOS less conceptually accessible and far less reliable will require developers to band together, thus developing the kind of community where strangers talk to one- another, kind of like at a grocery store before a huge snowstorm.

In the Beginning . . .

Following this same program, it will likely be necessary to move the BeOS headquarters to a far-less-comfortable climate. General environmental discomfort will breed this attitude within and there truly is no greater recipe for success. I would suggest Seattle, but I think it's already taken. You might try Washington, DC, but definitely not somewhere like San Diego or Tucson.

Unfortunately, the Be bug reporting system strips off the names of the people who report the bugs (to protect them from retribution!?) therefore I don't know who wrote this.

So it would appear that I'm in the middle of crowing about the technical and moral superiority of Debian Linux. But as almost always happens in the OS world, it's more complicated than that. I have Windows NT running on another machine, and the other day (Jan. 1999), when I had a problem with it, I decided to have another go at Microsoft Support. This time the search engine actually worked (though in order to reach it I had to identify myself as "advanced"). And instead of coughing up some useless FAQ, it located about two hundred documents (I was using very vague search criteria) that were obviously bug reports—though they were called something else. Microsoft, in other words, has actually got a system up and running that is functionally equivalent to Debian's bug database. It looks and feels different, of course, and it took a long time for me to find it, but it contains technical nitty-gritty and makes no bones about the existence of errors.

As I've explained, selling OSes for money is a basically untenable position, and the only way Apple and Microsoft can get away with it is by pursuing technological advancements as aggressively as they can, and by getting people to believe in, and to pay for, a particular image: in the case of Apple, that of the creative free thinker, and in the case of Microsoft, that of the respectable techno-bourgeois. Just like Disney, they're making money from selling an interface, a magic mirror. It has to be polished and seamless, or else the whole illusion is ruined and the business plan vanishes like a mirage.

Accordingly, it was the case until recently that the people who wrote manuals and created customer support websites for commercial OSes seemed to have been barred, by their employers' legal or PR departments, from admitting, even obliquely, that the software might contain bugs or that the interface might be suffering from the blinking twelve problem. They couldn't address users' actual difficulties. The manuals and websites were therefore useless, and caused even technically self-assured users to wonder whether they were going subtly insane.

When Apple engages in this sort of corporate behavior, one wants to believe that they are really trying their best. We all want to give Apple the benefit of the doubt, because mean old Bill Gates kicked the crap out of them, and because they have good PR. But when Microsoft does it, one almost cannot help becoming a paranoid conspiracist. Obviously they are hiding something from us! And yet they are so powerful! They are trying to drive us crazy!

This approach to dealing with one's customers was straight out of the Central European totalitarianism of the mid-twentieth century. The adjectives "Kafkaesque" and "Orwellian" come to mind. It couldn't last, any more than the Berlin Wall could, and so now Microsoft has a publicly available bug database. It's called something else, and it takes a while to find it, but it's there.

They have, in other words, adapted to the two-tiered Eloi/Morlock structure of technological society. If you're an Eloi you install Windows, follow the instructions, hope for the best, and dumbly suffer when it breaks. If you're a Morlock, you go to the website, tell it that you are "advanced," find the bug database, and get the truth straight from some anonymous Microsoft engineer.

But once Microsoft has taken this step, it raises the question, once again, of whether there is any point to being in the OS business at all. Customers might be willing to pay $95 to report a problem to Microsoft if, in return, they get some advice that no other user is getting. This has the useful side effect of keeping the users alienated from one another, which helps maintain the illusion that bugs are rare aberrations. But once the results of those bug reports become openly available on the Microsoft website, everything changes. No one is going to cough up $95 to report a problem when chances are good that some other sucker will do it first, and that instructions on how to fix the bug will then show up, for free, on a public site. And as the size of the bug database grows, it eventually becomes an open admission, on Microsoft's part, that their OS has just as many bugs as their competitors'. There is no shame in that; but it puts Microsoft on an equal footing with the others and makes it a lot harder for their customers—who Want to Believe—to believe.

MEMENTO MORI

Once the Linux machine has finished spitting out its jargonic opening telegram, it prompts me to log in with a user name and a password. At this point the machine is still running the command line interface, with white letters on a black screen. There are no windows, menus, or buttons. It does not respond to the mouse; it doesn't even know that the mouse is there. It is still possible to run a lot of software at this point. Emacs, for example, exists in both a CLI and a GUI version (actually there are two GUI versions, reflecting some sort of doctrinal schism between Richard Stallman and some other hackers). The same is true of many other Unix programs. Many don't have a GUI at all, and many that do are capable of running from the command line.

Of course, since my computer only has one monitor, I can only see one command line, and so you might think that I could only interact with one program at a time. But if I hold down the Alt key and then hit the F2 function button at the top of my keyboard, I am presented with a fresh, blank, black screen with a login prompt at the top of it. I can log in here and start some other program, then hit Alt-F1 and go back to the first screen, which is still doing whatever it was when I left it. Or I can do Alt-F3 and log in to a third screen, or a fourth, or a fifth. On one of these screens I might be logged in as myself, on another as root (the system administrator), on yet another I might be logged on to some other computer over the Internet.

Each of these screens is called, in Unix-speak, a tty, which is an abbreviation for "teletype." So when I use my Linux system in this way, I am going right back to that small room at Ames High School where I first wrote code twenty-five years ago, except that a tty is quieter and faster than a teletype, and capable of running vastly superior software, such as emacs or the GNU development tools.

It is easy (easy by Unix, not Apple/Microsoft standards) to configure a Linux machine so that it will go directly into a GUI when you boot it up. This way, you never see a tty screen at all. I still have mine boot into the white-on-black teletype screen however, as a computational memento mori. It used to be fashionable for a writer to keep a human skull on his desk as a reminder that he was mortal, that all about him was vanity. The tty screen reminds me that the same thing is true of slick user interfaces.

The XWindows system, which is the GUI of Unix, has to be capable of running on hundreds of different video cards with different chipsets, amounts of onboard memory, and motherboard buses. Likewise, there are hundreds of different types of monitors on the new and used market, each with different specifications, and so there are probably upwards of a million different possible combinations of card and monitor. The only thing they all have in common is that they all work in VGA mode, which is the old command line screen that you see for a few seconds when you launch Windows. So Linux always starts in VGA, with a teletype interface, because at first it has no idea what sort of hardware is attached to your computer. In order to get beyond the glass teletype and into the GUI, you have to tell Linux exactly what kinds of hardware you have. If you get it wrong, you'll get a blank screen at best, and at worst you might actually destroy your monitor by feeding it signals it can't handle.

When I started using Linux this had to be done by hand. I once spent the better part of a month trying to get an oddball monitor to work for me, and filled the better part of a composition book with increasingly desperate scrawled notes. Nowadays, most Linux distributions ship with a program that automatically scans the video card and self-configures the system, so getting XWindows up and running is nearly as easy as installing an Apple/Microsoft GUI. The crucial information goes into a file (an ASCII text file, naturally) called XF86Config, which is worth looking at even if your distribution creates it for you automatically. For most people it looks like meaningless cryptic incantations, which is the whole point of looking at it. An Apple/Microsoft system needs to have the same information in order to launch its GUI, but it's apt to be deeply hidden somewhere, and it's probably in a file that can't even be opened and read by a text editor. All of the important files that make Linux systems work are right out in the open. They are always ASCII text files, so you don't need special tools to read them. You can look at them any time you want, which is good, but you can also mess them up and render your system totally dysfunctional, which is not so good.

At any rate, assuming that my XF86Config file is just so, I enter the command "startx" to launch the XWindows system. The screen blanks out for a minute, the monitor makes strange twitching noises, then reconstitutes itself as a blank gray desktop with a mouse cursor in the middle. At the same time it is launching a window manager. XWindows is pretty low-level software; it provides the infrastructure for a GUI, and it's a heavy industrial infrastructure. But it doesn't do windows. That's handled by another category of application that sits atop XWindows, called a window manager. Several of these are available, all free of course. The classic is twm (Tom's Window Manager), but there is a smaller and supposedly more efficient variant of it called fvwm, which is what I use. I have my eye on a completely different window manager called Enlightenment, which may be the hippest single technology product I have ever seen, in that (a) it is for Linux, (b) it is freeware, (c) it is being developed by a very small number of obsessed hackers, and (d) it looks amazingly cool; it is the sort of window manager that might show up in the backdrop of an *Alien* movie.

Anyway, the window manager acts as an intermediary between XWindows and whatever software you want to use. It draws the window frames, menus, and so on, while the applications themselves draw the actual content in the windows. The applications might be of any sort: text editors, web browsers, graphics packages, or utility programs—such as a clock or calculator. In other words, from this point on, you feel as if you have been shunted into a parallel universe that is quite similar to the familiar Apple or Microsoft one, but slightly and pervasively different. The premier graphics program under Apple/Microsoft is Adobe Photoshop, but under Linux it's something called the GIMP. Instead of the Microsoft Office Suite, you can buy something called ApplixWare. Many commercial software packages, such as Mathematica, Netscape Communicator, and Adobe Acrobat, are available in Linux versions, and depending on how you set up your window manager, you can make them look and behave just as they would under MacOS or Windows.

But there is one type of window you'll see on Linux GUI that is rare or nonexistent under other OSes. These windows are called "xterm" and contain nothing but lines of text—this time, black text on a white background, though you can make them be different colors if you choose. Each xterm window is a separate command line interface—a tty in a window. So even when you are in full GUI mode, you can still talk to your Linux machine through a command line interface.

There are many good pieces of Unix software that do

not have GUIs at all. This might be because they were developed before XWindows was available, or because the people who wrote them did not want to suffer through all the hassle of creating a GUI, or because they simply do not need one. In any event, those programs can be invoked by typing their names into the command line of an xterm window. The whoami command, mentioned earlier, is a good example. There is another called wc ("word count"), which simply returns the number of lines, words, and characters in a text file.

The ability to run these little utility programs on the command line is a great virtue of Unix, and one that is unlikely to be duplicated by pure GUI operating systems. The wc command, for example, is the sort of thing that is easy to write with a command line interface. It probably does not consist of more than a few lines of code, and a clever programmer could probably write it in a single line. In compiled form it takes up just a few bytes of disk space. But the code required to give the same program a graphical user interface would probably run into hundreds or even thousands of lines, depending on how fancy the programmer wanted to make it. Compiled into a runnable piece of software, it would have a large overhead of GUI code. It would be slow to launch and it would use up a lot of memory. This would simply not be worth the effort, and so wc would never be written as an independent program at all. Instead users would have to wait for a word count feature to appear in a commercial software package.

GUIs tend to impose a large overhead on every single

piece of software, even the smallest, and this overhead completely changes the programming environment. Small utility programs are no longer worth writing. Their functions, instead, tend to get swallowed up into omnibus software packages. As GUIs get more complex, and impose more and more overhead, this tendency becomes more pervasive, and the software packages grow ever more colossal. After a point they begin to merge with each other, as Microsoft Word and Excel and Power-Point have merged into Microsoft Office: a stupendous software Wal-Mart sitting on the edge of a town filled with tiny shops that are all boarded up.

It is an unfair analogy, because when a tiny shop gets boarded up it means that some small shopkeeper has lost his business. Of course nothing of the kind happens when wc becomes subsumed into one of Microsoft Word's countless menu items. The only real drawback is a loss of flexibility for the user, but it is a loss that most customers obviously do not notice or care about. The most serious drawback to the Wal-Mart approach is that most users only want or need a tiny fraction of what is contained in these giant software packages. The remainder is clutter, dead weight. And yet the user in the next cubicle over will have completely different opinions as to what is useful and what isn't.

The other important thing to mention, here, is that Microsoft has included a genuinely cool feature in the Office package: a Visual Basic programming package. Basic is the first computer language that I learned, back when I was using the paper tape and the teletype. By using Visual Basic—a modernized version of the language that comes with Office—you can write your own little utility programs that know how to interact with all of the little doohickeys, gewgaws, bells, and whistles in Office. Basic is easier to use than the languages typically employed in Unix command line programming, and Office has reached many, many more people than the GNU tools. And so it is quite possible that this feature of Office will, in the end, spawn more hacking than GNU.

But now I'm talking about application software, not operating systems. And as I've said, Microsoft's application software tends to be very good stuff. I don't use it very much, because I am nowhere near their target market. If Microsoft ever makes a software package that I use and like, then it really will be time to dump their stock, because I am a market segment of one.

GEEK FATIGUE

Over the years that I've been working with Linux I have
filled three and a half notebooks logging my experiences.
I only begin writing things down when I'm doing something complicated, like setting up XWindows or fooling
around with my Internet connection, and so these notebooks contain only the record of my struggles and frus-
trations. When things are going well for me, I'll work
along happily for many months without jotting down a
single note. So these notebooks make for pretty bleak
reading. Changing anything under Linux is a matter of
opening up various of those little ASCII text files and
changing a word here and a character there, in ways that
are extremely significant to how the system operates.

Many of the files that control how Linux operates are
nothing more than command lines that became so long
and complicated that not even Linux hackers could type
them correctly. When working with something as powerful as Linux, you can easily devote a full half-hour to
engineering a single command line. For example, the
"find" command, which searches your file system for files

that match certain criteria, is fantastically powerful and general. Its "man" is eleven pages long, and these are pithy pages; you could easily expand them into a whole book. And if that is not complicated enough in and of itself, you can always pipe the output of one Unix command to the input of another, equally complicated one. The "pon" command, which is used to fire up a PPP connection to the Internet, requires so much detailed information that it is basically impossible to launch it entirely from the command line. Instead, you abstract big chunks of its input into three or four different files. You need a dialing script, which is effectively a little program telling it how to dial the phone and respond to various events; an options file, which lists up to about sixty different options on how the PPP connection is to be set up; and a secrets file, giving information about your password.

Presumably there are godlike Unix hackers somewhere in the world who don't need to use these little scripts and options files as crutches, and who can simply pound out fantastically complex command lines without making typographical errors and without having to spend hours flipping through documentation. But I'm not one of them. Like almost all Linux users, I depend on having all of those details hidden away in thousands of little ASCII text files, which are in turn wedged into the recesses of the Unix file system. When I want to change something about the way my system works, I edit those files. I know that if I don't keep track of every little change I've made, I won't be able to get the system back in working order after I've gotten it all messed up. Keeping handwritten logs is tedious, not to mention kind of anachronistic. But it's necessary.

I probably could have saved myself a lot of headaches by doing business with a company called Cygnus Support, which exists to provide assistance to users of free software. But I didn't, because I wanted to see if I could do it myself. The answer turned out to be yes, but just barely. There are many tweaks and optimizations that I could probably make in my system that I have never gotten around to attempting, partly because I get tired of being a Morlock some days, and partly because I am afraid of fouling up a system that generally works well.

Though Linux works for me and many other users, its sheer power and generality is its Achilles' heel. If you know what you are doing, you can buy a cheap PC from any computer store, throw away the Windows disks that come with it, turn it into a Linux system of mind-boggling complexity and power. You can hook it up to twelve other Linux boxes and make it into part of a parallel computer. You can configure it so that a hundred different people can be logged on to it at once over the Internet, via as many modem lines, Ethernet cards, TCP/IP sockets, and packet radio links. You can hang half a dozen different monitors off of it and play Doom with someone in Australia while tracking communications satellites in orbit and controlling your house's lights and thermostats and streaming live video from your webcam and surfing the Net and designing circuit boards on the other screens. But the sheer power and complexity of the system—the qualities that make it so vastly technically superior to other OSes—sometimes make it seem too formidable for routine day-to-day use.

Sometimes, in other words, I just want to go to Disneyland.

The ideal OS for me would be one that had a well-designed GUI that was easy to set up and use, but that included terminal windows where I could revert to the command line interface, and run GNU software, when it made sense. A few years ago, Be Inc. invented exactly that OS. It is called the BeOS.

ETRE

Many people in the computer business have had a difficult time grappling with Be, Incorporated, for the simple reason that nothing about it seems to make any sense whatsoever. It was launched in late 1990, which makes it roughly contemporary with Linux. From the beginning it has been devoted to creating a new operating system that is, by design, incompatible with all the others (though, as we shall see, it is compatible with Unix in some very important ways). If a definition of "celebrity" is someone who is famous for being famous, then Be is an anticelebrity. It is famous for not being famous; it is famous for being doomed. But it has been doomed for an awfully long time.

Be's mission might make more sense to hackers than to other people. In order to explain why, I need to explain the concept of "cruft," which, to people who write code, is nearly as abhorrent as unnecessary repetition.

If you've been to San Francisco, you may have seen older buildings that have undergone "seismic upgrades," which frequently means that grotesque superstructures of modern steelwork are erected around buildings made in, say, a classical style. When new threats arrive—if we have an Ice Age, for example—additional layers of even more high-tech stuff may be constructed, in turn, around these, until the original building is like a holy relic in a cathedral—a shard of yellowed bone enshrined in tons of fancy protective junk.

Analogous measures can be taken to keep creaky old operating systems working. It happens all the time. Ditching a worn-out old OS ought to be simplified by the fact that, unlike old buildings, OSes have no aesthetic or cultural merit that makes them intrinsically worth saving. But it doesn't work that way in practice. If you work with a computer, you have probably customized your "desktop," the environment in which you sit down to work every day, and spent a lot of money on software that works in that environment, and devoted much time to familiarizing yourself with how it all works. This takes a lot of time, and time is money. As already mentioned, the desire to have one's interactions with complex technologies simplified through the interface, and to surround yourself with virtual tchotchkes and lawn ornaments, is natural and pervasive—presumably a reaction against the complexity and formidable abstraction of the computer world. Computers give us more choices than we really want. We prefer to make those choices once, or accept the defaults handed to us by software companies, and let sleeping dogs lie. But when an OS gets changed, all the dogs jump up and start barking.

The average computer user is a technological antiquarian who doesn't really like things to change. He or she is like an urban professional who has just bought a charming fixer-upper and is now moving the furniture and knicknacks around, and reorganizing the kitchen cupboards, so that everything's just right. If it is necessary for a bunch of engineers to scurry around in the basement, shoring up the foundation so that it can support the new cast-iron claw-foot bathtub, and snaking new wires and pipes through the walls to supply modern appliances, why, so be it—engineers are cheap, at least when millions of OS users split the cost of their services.

Likewise, computer users want to have the latest Pentium in their machines, and to be able to surf the web, without messing up all the stuff that makes them feel as if they know what the hell is going on. Sometimes this is actually possible. Adding more RAM to your system is a good example of an upgrade that is not likely to screw anything up.

Alas, very few upgrades are this clean and simple. Lawrence Lessig, the whilom Special Master in the Justice Department's antitrust suit against Microsoft, complained that he had installed Microsoft Internet Explorer on his computer, and in so doing, lost all of his bookmarks—his personal list of signposts that he used to navigate through the maze of the Internet. It was as if he'd bought a new set of tires for his car, and then, when pulling away from the garage, discovered that, owing to some inscrutable side effect, every signpost and road map in the world had been destroyed. If he's like most of us, he had put a lot of work into compiling that list of bookmarks. This is only a small taste of the sort of trouble that upgrades can cause. Crappy old OSes have value in the basically negative sense that changing to new ones makes us wish we'd never been born.

All of the fixing and patching that engineers must do in order to give us the benefits of new technology without forcing us to think about it, or to change our ways, produces a lot of code that, over time, turns into a giant clot of bubble gum, spackle, baling wire, and duct tape surrounding every operating system. In the jargon of hackers, it is called "cruft." An operating system that has many, many layers of cruft is described as "crufty." Hackers hate to do things twice, but when they see something crufty, their first impulse is to rip it out, throw it away, and start anew.

If Mark Twain were brought back to San Francisco today and dropped into one of these old seismically upgraded buildings, it would look just the same to him, with all the doors and windows in the same places—but if he stepped outside, he wouldn't recognize it. And—if he'd been brought back with his wits intact—he might question whether the building had been worth going to so much trouble to save. At some point, one must ask the question: Is this really worth it, or should we maybe just tear it down and put up a good one? Should we throw another human wave of structural engineers at stabilizing the Leaning Tower of Pisa, or should we just let the damn thing fall over and build a tower that doesn't suck?

Like an upgrade to an old building, cruft always seems

like a good idea when the first layers of it go on—just routine maintenance, sound prudent management. This is especially true if (as it were) you never look into the cellar, or behind the drywall. But if you are a hacker who spends all his time looking at it from that point of view, cruft is fundamentally disgusting, and you can't avoid wanting to go after it with a crowbar. Or, better yet, simply walk out of the building—let the Leaning Tower of Pisa fall over—and go make a new one *that doesn't lean.*

For a long time it was obvious to Apple, Microsoft, and their customers that the first generation of GUI operating systems was doomed, and that they would eventually need to be ditched and replaced with completely fresh ones. During the late eighties and early nineties, Apple launched a few abortive efforts to make fundamentally new post-Mac OSes, such as Pink and Taligent. When those efforts failed, they launched a new project called Copland—which also failed. In 1997 they flirted with the idea of acquiring Be, but instead they acquired Next, which has an OS called NextStep, which is, in effect, another variant of Unix. As these efforts went on, and on, and on, and failed and failed and failed, Apple's engineers, who were among the best in the business, kept layering on the cruft. They were gamely trying to turn the little toaster into a multitasking, Internet-savvy machine, and did an amazingly good job of it for a while—sort of like a movie hero running across a jungle river by hopping across crocodiles' backs. But in the real world you eventually run out of crocodiles, or step on a really smart one.

Speaking of which, Microsoft tackled the same problem in a considerably more orderly way by creating a new OS called Windows NT, which is explicitly intended to be a direct competitor of Unix. NT stands for "New Technology," which might be read as an explicit rejection of cruft. And indeed, NT is reputed to be a lot less crufty than what MacOS eventually turned into; at one point the documentation needed to write code on the Mac filled something like twenty-four binders. Windows 95 was, and Windows 98 is, crufty because they have to be backward-compatible with older Microsoft OSes. Linux deals with the cruft problem in the same way that, according to the tales we used to be told in school, Eskimos supposedly dealt with senior citizens: if you insist on using old versions of Linux software, you will sooner or later find yourself drifting through the Bering Straits on a dwindling ice floe. They can get away with this because most of the software is free, so it costs nothing to download up-to-date versions, and because most Linux users are Morlocks.

The great idea behind BeOS was to start from a clean sheet of paper and design an OS the right way. And that is exactly what they did. This was obviously a good idea from an aesthetic standpoint, but does not a sound business plan make. Some people I know in the GNU/Linux world are annoyed with Be for going off on this quixotic adventure when their formidable skills could have been put to work helping to promulgate Linux.

Indeed, none of it makes sense until you remember that the founder of the company, Jean-Louis Gassee, is from France—a country that for many years maintained its own separate and independent version of the English monarchy at a court in St. Germaines, complete with courtiers, coronation ceremonies, a state religion, and a foreign policy. Now, the same annoying yet admirable stiff-neckedness that gave us the Jacobites, the *force de frappe*, Airbus, and *Arret* signs in Quebec has brought us a really cool operating system. I fart in your general direction, Anglo-Saxon pig-dogs!

To create an entirely new OS from scratch, just because none of the existing ones was exactly right, struck me as an act of such colossal nerve that I felt compelled to support it. I bought a BeBox as soon as I could. The BeBox was a dual-processor machine, powered by Motorola chips, made specifically to run the BeOS; it could not run any other operating system. That's why I bought it. I felt it was a way to burn my bridges. Its most distinctive feature is two columns of LEDs on the front panel that zip up and down like tachometers to convey a sense of how hard each processor is working. I thought it looked cool, and besides, I reckoned that when the company went out of business in a few months, my BeBox would be a valuable collector's item.

Now it is about two years later and I am typing this on my BeBox. The LEDs (Das Blinkenlights, as they are called in the Be community) flash merrily next to my right elbow as I hit the keys. Be, Inc. is still in business, though they stopped making BeBoxes almost immediately after I bought mine. They made the sad, but probably quite wise, decision that hardware was a sucker's game, and ported the BeOS to Macintoshes and Mac clones. Since these used the same sort of Motorola chips that powered the BeBox, this wasn't especially hard.

Very soon afterwards, Apple strangled the Mac-clone makers and restored its hardware monopoly. So, for a while, the only new machines that could run BeOS were made by Apple.

By this point Be, like Spiderman with his Spider-sense, had developed a keen sense of when they were about to get crushed like a bug. Even if they hadn't, the notion of being dependent on Apple—so frail and yet so vicious—for their continued existence should have put a fright into anyone. Now engaged in their own crocodile-hopping adventure, they ported the BeOS to Intel chips—the same chips used in Windows machines. And not a moment too soon, for when Apple came out with its new top-of-the-line hardware, based on the Motorola G3 chip, they withheld the technical data that Be's engineers would need to make the BeOS run on those machines. This would have killed Be, just like a slug between the eyes, if they hadn't already made the jump to Intel.

So now BeOS runs on an assortment of hardware that is almost incredibly motley: BeBoxes, aging Macs and Mac orphan-clones, and Intel machines that are intended to be used for Windows. Of course the latter type are ubiquitous and shockingly cheap nowadays, so it would appear that Be's hardware troubles are finally over.

Some German hackers have even come up with a Das Blinkenlights replacement: it's a circuit board kit that you can plug into PC-compatible machines running BeOS to give you the zooming LED tachometers that were such a popular feature of the BeBox.

My BeBox is already showing its age, as all computers do after a couple of years, and sooner or later I'll probably have to replace it with an Intel machine. Even after that, though, I will still be able to use it. Because, inevitably, someone has now ported Linux to the BeBox.

At any rate, BeOS has an extremely well-thought-out GUI built on a technological framework that is solid. It is based from the ground up on modern object-oriented software principles. BeOS software consists of quasi-independent software entities called objects, which communicate by sending messages to each other. The OS itself is made up of such objects and serves as a kind of post office or Internet that routes messages to and fro, from object to object. The OS is multithreaded, which means that like all other modern OSes it can walk and chew gum at the same time, but it gives programmers a lot of power over spawning and terminating threads, or independent subprocesses. It is also a multiprocessing OS, which means that it is inherently good at running on computers that have more than one CPU (Linux and Windows NT can also do this proficiently).

For this user, a big selling point of BeOS is the built-in Terminal application, which enables you to open up windows that are equivalent to the xterm windows in Linux. In other words, the command line interface is available, if you want it. And because BeOS hews to a certain standard called POSIX, it is capable of running most of the GNU software. That is to say that the vast array of command line software developed by the GNU crowd will work in BeOS terminal windows without complaint. This includes the GNU development tools— the compiler and linker. And it includes all of the handy little utility programs. I'm writing this using a modern sort of user-friendly text editor called Pe, written by a Dutchman named Maarten Hekkelman. When I want to find out how long my essay is, I jump to a terminal window and run wc.

As is suggested by the sample bug report I quoted earlier, people who work for Be and developers who write code for BeOS seem to be enjoying themselves more than their counterparts who work with other OSes. They also seem to be a more diverse lot in general. A couple of years ago I went to an auditorium at a local university to see some representatives of Be put on a dog-and-pony show. I went because I assumed that the place would be empty and echoing, and I felt that they deserved an audience of at least one. In fact, I ended up standing in an aisle, for hundreds of students had packed the place. It was like a rock concert. One of the two Be engineers on the stage was a black man, which unfortunately is a very odd thing in the high-tech world. The other made a ringing denunciation of cruft, and extolled BeOS for its cruft-free qualities, and actually came out and said that in ten or fifteen years, when BeOS had become all crufty like MacOS and Windows 95, it would be time to simply throw it away and create a new OS from scratch. I doubt that this is an official Be, Inc. policy, but it sure made a big impression on everyone in the room! During the late eighties, the MacOS was, for a time, the OS of cool people—artists and creative-minded hackers—and BeOS seems to have the potential to attract the same crowd now. Be mailing lists are crowded with hackers with names like Vladimir and Olaf and Pierre, sending flames to each other in fractured techno-English.

The only real question about BeOS is whether or not it is doomed.

Of late, Be has responded to the tiresome accusation that they are doomed with the assertion that BeOS is "a media operating system" made for media content creators, and hence is not really in competition with Microsoft Windows at all. This is a little bit disingenuous. To go back to the car dealership analogy, it is like the Batmobile dealer claiming that he is not really in competition with the others because his car can go three times as fast as theirs and is also capable of flying.

Be has an office in Paris and, as mentioned, the conversation on Be mailing lists has a strongly European flavor. At the same time they have made strenuous efforts to find a niche in Japan, and Hitachi has recently begun bundling BeOS with their PCs. So if I had to make a wild guess, I'd say that they are playing Go while Microsoft is playing chess. They are staying clear, for now, of Microsoft's overwhelmingly strong position in North America. They are trying to get themselves established around the edges of the board, as it were, in Europe and Japan, where people may be more open to alternative OSes, or at least more hostile to Microsoft, than they are in the United States.

What holds Be back in this country is that the smart people are afraid to look like suckers. You run the risk of looking naive when you say, "I've tried the BeOS and here's what I think of it." It seems much more sophisticated to say, "Be's chances of carving out a new niche in the highly competitive OS market are close to nil."

It is, in techno-speak, a problem of mindshare. And in the OS business, mindshare is more than just a PR issue; it has direct effects on the technology itself. All of the peripheral gizmos that can be hung off of a personal computer—the printers, scanners, PalmPilot interfaces, and Lego Mindstorms—require pieces of software called drivers. Likewise, video cards and (to a lesser extent) monitors need drivers. Even the different types of motherboards on the market relate to the OS in different ways, and separate code is required for each one. All of this hardware-specific code must not only be written but also tested, debugged, upgraded, maintained, and supported. Because the hardware market has become so vast and complicated, what really determines an OS's fate is not how good the OS is technically, or how much it costs, but rather the availability of hardware-specific code. Linux hackers have to write that code themselves, and they have done an amazingly good job of keeping up to speed. Be, Inc. has to write all their own drivers; though as BeOS has begun gathering momentum, third-party developers have begun to contribute drivers, which are available on Be's website.

But Microsoft owns the high ground at the moment, because it doesn't have to write its own drivers. Any hardware maker bringing a new video card or peripheral device to market today knows that it will be unsalable unless it comes with the hardware-specific code that will make it work under Windows, and so each hardware maker has accepted the burden of creating and maintaining its own library of drivers.

MINDSHARE

The U.S. government's assertion that Microsoft has a monopoly in the OS market might be the most patently absurd claim ever advanced by the legal mind. Linux, a technically superior operating system, is being given away for free, and BeOS is available at a nominal price. This is simply a fact, which has to be accepted whether or not you like Microsoft. Microsoft is really big and rich, and if some of the government's witnesses are to be believed, they are not nice guys. But the accusation of a monopoly simply does not make any sense.

What is really going on is that Microsoft has seized, for the time being, a certain type of high ground: they dominate in the competition for mindshare, and so any hardware or software maker who wants to be taken seriously feels compelled to make a product that is compatible with their operating systems. Since Windows-compatible drivers get written by the hardware makers, Microsoft doesn't have to write them; in effect, the hardware makers are adding new components to Windows, making it a more capable OS, without charging Microsoft for the service. It is a very good position to be in. The only way to fight such an opponent is to have an army of highly competetent coders who write and distribute equivalent drivers, which Linux does.

But possession of this psychological high ground is different from a monopoly in any normal sense of that word, because here the dominance has nothing to do with technical performance or price. The old robber-baron monopolies were monopolies because they physically controlled means of production and/or distribution. But in the software business, the means of production is hackers typing code, and the means of distribution is the Internet, and no one is claiming that Microsoft controls those.

Here, instead, the dominance is inside the minds of people who buy software. Microsoft has power because people believe it does. This power is very real. It makes lots of money. Judging from recent legal proceedings in both Washingtons, it would appear that this power and this money have inspired some very peculiar executives to come out and work for Microsoft, and that Bill Gates should have administered saliva tests to some of them before issuing them Microsoft ID cards.

But this is not the sort of power that fits any normal definition of the word "monopoly," and it's not amenable to a legal fix. The courts may order Microsoft to do things differently. They might even split the company up. But they can't really do anything about a mindshare monopoly, short of taking every man, woman, and child in the developed world and subjecting them to a lengthy brainwashing procedure.

Mindshare dominance is, in other words, a really odd sort of beast, something that the framers of our antitrust laws couldn't possibly have imagined. It looks like one of these modern, wacky chaos-theory phenomena, a complexity thing, in which a whole lot of independent but connected entities (the world's computer users), making decisions on their own, according to a few simple rules of thumb, generate a large phenomenon (total domination of the market by one company) that cannot be made sense of through any kind of rational analysis. Such phenomena are fraught with concealed tipping-points and all a-tangle with bizarre feedback loops, and cannot be understood; people who try, end up going crazy, forming crackpot theories, or becoming high-paid chaos-theory consultants.

Now, there might be one or two people at Microsoft who are dense enough to believe that mindshare dominance is some kind of stable and enduring position. Maybe that even accounts for some of the weirdos they've hired in the pure business end of the operation, the zealots who keep getting hauled into court by enraged judges. But most of them must have the wit to understand that phenomena like these are maddeningly unstable, and that there's no telling what weird, seemingly inconsequential event might cause the system to shift into a radically different configuration.

To put it another way, Microsoft can be confident that Thomas Penfield Jackson will not hand down an order that the brains of everyone in the developed world are to be summarily reprogrammed. But there's no way to predict when people will decide, en masse, to reprogram their own brains. This might explain some of Microsoft's behavior, such as their policy of keeping eerily large reserves of cash sitting around, and the extreme anxiety that they display whenever something like Java comes along.

I have never seen the inside of the building at Microsoft where the top executives hang out, but I have this fantasy that in the hallways, at regular intervals, big red alarm boxes are bolted to the wall. Each contains a large red button protected by a windowpane. A metal hammer dangles on a chain next to it. Above is a big sign reading: IN THE EVENT OF A CRASH IN MARKET SHARE, BREAK GLASS.

What happens when someone shatters the glass and hits the button, I don't know, but it sure would be interesting to find out. One imagines banks collapsing all over the world as Microsoft withdraws its cash reserves, and shrink-wrapped pallet-loads of hundred-dollar bills dropping from the skies. No doubt, Microsoft has a plan. But what I would really like to know is whether, at some level, their programmers might heave a big sigh of relief if the burden of writing the One Universal Interface to Everything were suddenly lifted from their shoulders.

THE RIGHT PINKY OF GOD

In his book *The Life of the Cosmos*, which everyone should read, Lee Smolin gives the best description I've ever read of how our universe emerged from an uncannily precise balancing of different fundamental constants. The mass of the proton, the strength of gravity, the range of the weak nuclear force, and a few dozen other fundamental constants completely determine what sort of universe will emerge from a Big Bang. If these values had been even slightly different, the universe would have been a vast ocean of tepid gas or a hot knot of plasma or some other basically uninteresting thing—a dud, in other words. The only way to get a universe that's not a dud—that has stars, heavy elements, planets, and life— is to get the basic numbers just right. If there were some machine, somewhere, that could spit out universes with randomly chosen values for their fundamental constants, then for every universe like ours it would produce 10^{229} duds.

Though I haven't sat down and run the numbers on it, to me this seems comparable to the probability of making a Unix computer do something useful by logging into a tty and typing in command lines when you have forgotten all of the little options and keywords. Every time your right pinky slams that Enter key, you are making another try. In some cases the operating system does nothing. In other cases it wipes out all of your files. In most cases it just gives you an error message. In other words, you get many duds. But sometimes, if you have it all just right, the computer grinds away for a while and then produces something like emacs. It actually generates complexity, which is Smolin's criterion for interestingness.

Not only that, but it's beginning to look as if, once you get below a certain size—way below the level of quarks, down into the realm of string theory—the universe can't be described very well by physics as it has been practiced since the days of Newton. If you look at a small enough scale, you see processes that look almost computational in nature.

I think that the message is very clear here: somewhere outside of and beyond our universe is an operating system, coded up over incalculable spans of time by some kind of hacker-demiurge. The cosmic operating system uses a command line interface. It runs on something like a teletype, with lots of noise and heat; punched-out bits flutter down into its hopper like drifting stars. The demiurge sits at his teletype, pounding out one command line after another, specifying the values of fundamental constants of physics:

universe -G 6.672e-11 -e 1.602e-19 -h 6.626e-34
-protonmass 1.673e-27. . . .

and when he's finished typing out the command line, his
right pinky hesitates above the enter key for an aeon or
two, wondering what's going to happen; then down it
comes—and the *whack* you hear is another Big Bang.

Now *that* is a cool operating system, and if such a thing
were actually made available on the Internet (for free,
of course), every hacker in the world would download it
right away and then stay up all night long messing with
it, spitting out universes right and left. Most of them
would be pretty dull universes, but some of them would
be simply amazing. Because what those hackers would
be aiming for would be much more ambitious than a
universe that had a few stars and galaxies in it. Any run-
of-the-mill hacker would be able to do that. No, the way
to gain a towering reputation on the Internet would be
to get so good at tweaking your command line that your
universes would spontaneously develop life. And once
the way to do that became common knowledge, those
hackers would move on, trying to make their universes
develop the right kind of life, trying to find the one
change in the nth decimal place of some physical con-
stant that would give us an earth in which, say, Hitler
had been accepted into art school after all.

Even if that fantasy came true, though, most users (including myself, on certain days) wouldn't want to bother
learning to use all of those arcane commands and strug-
gling with all of the failures; a few dud universes can

really clutter up your basement. After we'd spent a while
pounding out command lines and hitting that enter key
and spawning dull, failed universes, we would start to
long for an OS that would go all the way to the opposite
extreme: an OS that had the power to do everything—
to live our life for us. In this OS, all of the possible
decisions we could ever want to make would have been
anticipated by clever programmers and condensed into
a series of dialog boxes. By clicking on radio buttons
we could choose from among mutually exclusive choices
(HETEROSEXUAL/HOMOSEXUAL). Columns of check boxes
would enable us to select the things that we wanted in
our life (GET MARRIED/WRITE GREAT AMERICAN NOVEL)
and for more complicated options we could fill in little
text boxes (NUMBER OF DAUGHTERS/NUMBER OF SONS:).

Even this user interface would begin to look awfully
complicated after a while, with so many choices and so
many hidden interactions between choices. It could become damn near unmanageable—the blinking twelve
problem all over again. The people who brought us this
operating system would have to provide templates and
wizards, giving us a few default lives that we could use
as starting places for designing our own. Chances are
that these default lives would actually look pretty good
to most people, good enough, anyway, that they'd be
reluctant to tear them open and mess around with them
for fear of making them worse. So after a few releases,
the software would begin to look even simpler: you
would boot it up and it would present you with a dialog
box with a single large button in the middle labeled:

LIVE. Once you had clicked that button, your life would begin. If anything got out of whack, or failed to meet your expectations, you could complain about it to Microsoft's Customer Support Department. If you got a flack on the line, he or she would tell you that your life was actually fine, that there was not a thing wrong with it, and in any event it would be a lot better after the next upgrade was rolled out. But if you persisted, and identified yourself as advanced, you might get through to an actual engineer.

What would the engineer say, after you had explained your problem and enumerated all of the dissatisfactions in your life? He would probably tell you that life is a very hard and complicated thing; that no interface can change that; that anyone who believes otherwise is a sucker; and that if you don't like having choices made for you, you should start making your own.

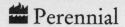 **Perennial**

Books by Neal Stephenson:

CRYPTONOMICON
ISBN 0-380-78862-4

With this extraordinary first volume in what promises to be an epoch-making masterpiece, Neal Stephenson hacks into the secret histories of nations and the private obsessions of men, decrypting with dazzling virtuosity the forces that shaped this century. A breathtaking tour-de-force and Neal Stephenson's most accomplished and affecting work to date, *Cryptonomicon* is profound and prophetic, hypnotic and hyper-driven, as it leaps forward and back between World War II and the World Wide Web, hinting all the while at a dark day-after-tomorrow. It is a work of great art, thought, and creative daring; the product of a true icon.

"Electrifying . . . hilarious . . . a sprawling, picaresque novel about code making and code breaking . . . . Stephenson cares as much about telling good stories as he does about farming out cool ideas."

—*The New York Times Book Review*

IN THE BEGINNING WAS THE COMMAND LINE
ISBN 0-380-81593-1

This is "the Word"—one man's word, certainly—about the art (and artifice) of the state of our computer-centric existence. Mostly well-reasoned examination and partial rant, this is an irreverent, hilarious treatise on the cyber-culture past and present; on operating system tyrannies and downloaded popular revolutions; on the Internet, Disney World, Big Bangs, not to mention the meaning of life itself.

"A powerful voice of the cyber age."—*USA Today*

THE BIG U
ISBN 0-380-81603-2

Now available again, Stephenson's first novel is a satire of an American university that is smart, and funny. Casimir Randon's introduction to American Megaversity is fraught with red tape, Newspeak, and enrollment procedures based on the catch-22 principle. Having struggled long and hard to afford a college education, Casimir has come up against the awful truth. What is he doing at the Big U?

"An entertaining and sometimes murderous satire on campus life."

—*The New York Times Book Review*

Available wherever books are sold, or call 1-800-331-3761 to order.